D0203471

The Transfer of Early Industrial Technologies to America

The Transfer of Early Industrial Technologies to America

Darwin H. Stapleton

American Philosophical Society
Independence Square Philadelphia
1987

MEMOIRS OF THE
AMERICAN PHILOSOPHICAL SOCIETY
Held at Philadelphia
For Promoting Useful Knowledge
Volume 177

Manufactured in the U.S.A. by Allen Press, Lawrence, Kansas

Library of Congress Catalog Card. No. 86-72882
International Standard Book No. 0-87169-177-9
US ISSN 0065-9738

Cover: Water-Power Blowing Engine, Catasauqua, Pennsylvania

To my Mother and Father, with love.

Preface

As I reflect on my reasons for writing this book, I realize that it is the product of a lifetime of interests and experience. When I was a child my father often pointed out to me the artifacts of nineteenth-century industry on the Pennsylvania landscape. Old iron furnace and gristmill ponds where we fished, traces of canals where we walked or stopped to read historical markers, and the restored Hopewell ironworks are indelible memories.

Though I subsequently wrote occasional papers on early American industry in school and at Swarthmore College, my studies were normally far removed from blast furnaces and canal locks. On entering the Hagley Graduate Program at the University of Delaware I first had the opportunity to immerse myself in the literature of industrialization and I first discovered the discipline of the history of technology.

It was under the tutelage of Eugene S. Ferguson that I was able to place my interest in early American industrialization within the context of technology transfer. As we discussed my dissertation topic I recognized that few writers had dealt with the European roots of American technologies, and fewer still with the process of transfer. Admiring the case-study approaches of Paul Strassman and Harold Passer, I decided to focus on particular examples of technology transfer.

I have since modified that earlier version of this study through further research on each case and by adding two chapters. An invitation to speak at the Lawrence Henry Gipson Symposium at Lehigh University provided the incentive and opportunity to write the chapter on Weston and Latrobe, utilizing materials I gathered during my decade as assistant editor for technology with The Papers of Benjamin Henry Latrobe. The survey in chapter 1 is substantially inspired by a number of studies published since I completed the original manuscript of this book.

In the course of research and writing Donna, my wife, has been my mainstay and steady critic in all phases of my work. Eugene S. Ferguson's search for the right questions shaped my approach even after he served as my immediate adviser. Those who have generously given time to read and criticize all or portions of the manuscript include John Beer, Reed Geiger, Carol Hoffecker, Robert Howard, John Rae, Alan Rocke, Tamara Stech, David J. Jeremy, John Van Horne, Norman Wilkinson, and an anonymous reviewer.

I have received so many services from libraries and archives that I cannot be certain of listing them all, but I particularly want to thank the following: American Philosophical Society Library, Hagley Library, Historical Society of Pennsylvania, Conrail (Reading Company), Swem Library of the College of William and Mary in Virginia, Virginia State Library, and William Penn Archives (Harrisburg, Pennsylvania).

Darwin H. Stapleton
March 1986

Contents

Illustrations

Part 1

A Survey

I: The Transfer of Industrial Technologies to Early America

The Historical Role of Technology Transfer

The transfer of technology has been a central factor in the process of industrialization wherever it has occurred. The British industrial revolution had many indigenous elements, but the influx of Huguenot and Flemish immigrants in the seventeenth century, and the judicious copying of various continental technologies in the eighteenth must be recognized if we are to understand fully the roots of much British technical innovation.[1]

Subsequently Britain became the source of many fundamental industrial technologies for continental nations and the United States, particularly in metallurgy, machine tools and machine building, and transport. In the nineteenth and throughout the twentieth century many non-Western nations consciously borrowed or had thrust upon them by colonial regimes the western European and American systems of technology which were thought to promote industrialization.[2] In the case of Japan the transfer of European technologies was

[1] A. E. Musson, "Continental Influences on the Industrial Revolution in Great Britain," in *Great Britain and Her World, 1750–1914: Essays in Honour of W. O. Henderson,* ed. Barrie M. Ratcliffe (Manchester: University of Manchester Press, 1975), 71–85.

[2] Examples from a flourishing literature are John H. Jensen and Gerhard Rosseger, "Transferring Technology to a Peripheral Economy: The Case of Lower Danube Transport Development, 1846–78," *Technology and Culture* 19 (October 1978): 675–702; W. O. Henderson, *Britain and Industrial Europe, 1750–1870* (Liverpool: Liverpool University Press, 1954); Eric Robinson, "The Transference of British Technology to Russia, 1760–1820: A Preliminary Inquiry," in Ratcliffe, ed., *Great Britain and Her World,* 1–26; Hans Rogger, "*Amerikanizm* and the Economic Development of Russia," *Comparative Studies in Society and History* 23 (July 1981): 382–420; William L. Blackwell, *The Beginnings of Russian Industrialization* (Princeton: Princeton University Press, 1968); David L. Jeremy, "British Textile Technology Transmission to the United States: The Philadelphia Region Experience, 1770–1820," *Business History Review* 47 (Spring 1973): 24–52.

extremely successful.[3] Moreover, reciprocal transfer of technology among the advanced industrial nations is now an accepted phenomenon of the modern world.

If the process of industrialization in the United States is to be fully understood, the transfer of crucial European industrial technologies must be examined. The ground for this endeavor has been cleared by demolishing the concept of "Yankee ingenuity" as an organizing principle for the history of American technology.[4] The new edifice will be established on the foundation of an appreciation of truly American innovations and recognition of the role of the transfer of technology.[5]

* * *

The industrialization of the United States in the nineteenth century occurred within the context of the basic European technologies brought by the colonists, and was spurred on by the innovative technologies brought from Europe after the American Revolution. Americans adapted (sometimes rejected) and diffused these technologies, establishing an industrial complex that ultimately overtook the originator of the industrial revolution in many areas.

As the term is most often used by historians, a transfer of technology is an attempt to take processes, work methods, and concepts from one nation or culture in order to establish them in another nation or culture. Transfers normally have involved skilled people migrating to a new land, or technically oriented persons learning new skills elsewhere and then bring-

[3] Akio Okochi and Hoshimi Uchida, eds., *Development and Diffusion of Technology: Electrical and Chemical Industries,* The International Conference on Business History, 6 (Tokyo: University of Tokyo Press, 1980), 125–223.

[4] John W. Oliver, *History of American Technology* (New York: Ronald, 1956), v, uses this concept. David A. Hounshell has argued that historians of American technology have been "unable to supplant . . . the contours of American technical history . . . defined by" such older works. David A. Hounshell, "On the Discipline of the History of American Technology," *Journal of American History* 67 (March 1981): 864.

[5] New foundations are indicated by such publications as Eugene S. Ferguson, "The American-ness of American Technology," *Technology and Culture* 20 (January 1979): 3–24; Carroll W. Pursell, ed., *Technology in America: A History of Individuals and Ideas* (Cambridge, Mass.: M.I.T. Press, 1981); Otto Mayr, ed., *Yankee Enterprise: The Rise of the American System of Manufactures* (Washington: Smithsonian Institution Press, 1982).

ing them back to their native soil. Over thirty years ago Warren C. Scoville suggested three useful categories into which such migrations may be classified: individuals, groups, and cultural minorities.[6]

Individual transfers have been the most common. This category included all those whose decision to move and choice of destination was limited to them or their family unit. Their skills may have been relatively self-sufficient, such as farming, or in the opposite extreme, may have required the support of a complex of other workers and investors to provide them suitable employment. (For example, a power-loom weaver worked in the context of a textile factory and could not ply his trade otherwise.) Individuals may have been the most likely to fail to transfer their skills if those skills were new and innovative, because such individuals sometimes found that they had little or no support from others during the trying period of introduction.

Group transfers occurred when several individuals in the same or related fields of technology chose to emigrate and settle together. Historically many group transfers were encouraged by government or private agencies which wished to establish a new industrial process. In many cases groups did not remain long at the specific site or enterprise for which they were recruited, but broke up and spread the new technology elsewhere.

Cultural minorities have emigrated by the hundreds or thousands under the pressures of war, famine, or social upheaval; and sometimes in response to the enticements or opportunities of a new land. Minority migrations sometimes had a geographic focus such that they became the dominant culture in a region. In some cases minority groups possessed such different technologies that they made important contributions to their new homeland. They also brought entire complexes of beliefs, values, and customs which formed the framework for their technologies and have helped insure that those technologies could survive a generation or more. Of course, groups and individuals also carried their beliefs, values, and customs,

[6] Warren C. Scoville, "Minority Migrations and the Diffusion of Technology," *Journal of Economic History* 9 (1951): 349–51.

but they were more readily compromised by the culture of the recipient society.

All three types of migration contributed to the transfer of technology from Europe to America prior to 1850. The seaboard colonies of North America, and later the United States, had a constant flow of emigrants, all of whom gave their skills to their new homeland, and some of whom made distinct transfers of industrial technology.

The Planting of European Technology in America

The European settlers of North America confronted a group of Indian tribes whose technology cannot properly be described as Neolithic, but which certainly lacked tools and processes associated with the rise of Western civilization such as wheeled transport, water mills, metal smelting, and sailing ships. Nonetheless, Indian technology was highly adapted to the North American landscape. The first generation of European settlers relied upon it heavily, and frontier settlements drew upon Indian practices into the nineteenth century.

Agriculture was at first undertaken in the Indian fashion (very successfully, it should be added), and Indian-style long houses and wigwams were often used for temporary accommodations. Canoes were for decades a fundamental form of settler transport, and deerskin clothing was adopted as an acceptable form of dress.[7]

The European technology that the settlers brought with them has been characterized as essentially "medieval,"[8] although elements of it foreshadowed the industrial age. Farmers relied upon the heavy wheeled horse- or ox-plow to break the

[7] Darret B. Rutman, *Winthrop's Boston: A Portrait of a Puritan Town, 1630–1649* (New York: Norton, 1972), 26, 36, 177; Charles F. Carroll, *The Timber Economy of New England* (Providence, R.I.: Brown University Press, 1973), 68; Francis Jennings, *The Invasion of America: Indians, Colonialism, and the Cant of Conquest* (New York: Norton, 1976), 33–34, 40–41, 62–63; Clarence H. Danhof, *Change in Agriculture: The Northern United States, 1820–1870* (Cambridge, Mass.: Harvard University Press, 1969), 49.

[8] Carroll W. Pursell, Jr., "Technology in America: An Introduction," in *Technology in America,* 1; Lynn White, Jr., *Medieval Religion and Technology: Collected Essays* (Berkeley: University of California Press, 1978), ch. 7, "The Legacy of the Middle Ages in the American Wild West."

soil and to make furrows for the seed. They sowed by hand "broadcasting," and harvested with scythes and sickles.

The transfer of this farming technology was not as straightforward as might be thought, primarily because the landscape and environment in America were often substantially different than in Europe and the soil and crops did not easily respond to standard European practices. Effective means of clearing land had to be learned, since there were trees to cut down, stumps to uproot, and boulders to remove, with rapid soil erosion a frequent consequence. Moreover, since most of the early immigrants were inexperienced at farming, much had to be learned through failure—particularly since the new crops of corn (maize) and tobacco became important so quickly.

Once, however, land was cleared and the staple crops of the colonies (corn, wheat, and tobacco) were established, farmers found European practices adequate. The most successful, by many accounts, were the Germans, who first settled heavily in Pennsylvania. Their agricultural patterns included deep plowing, the erection of sturdy farm buildings, and mixed livestock–grain farming. They used the land intensively, but were careful to restore its fertility by crop rotation and manuring.[9] The Pennsylvania Germans' style of farming was perhaps inappropriate for frontier regions, but in the settled regions of the East they were model farmers. Advocates of agricultural improvement in the later eighteenth century cited their practices as the ideal standard.[10]

The colonial craftsman stood on this agricultural base. A variety of craftsmen arrived at the first settlements, although severe pressures for food production, the small markets for craft products, and the competition of imports from England

[9] Richard H. Shyrock, "British Versus German Traditions in Colonial Agriculture," *Mississippi Valley Historical Review* 26 (June 1939): 46, 49–51.

[10] Lucius F. Ellsworth, "The Philadelphia Society for the Promotion of Agriculture and Agricultural Reform, 1785–1793," *Agricultural History* 42 (July 1968): 196, 198.

If in general the settler's agriculture was vitalized by absorbing Indian crops and techniques, and was sustained by the continued infusion of European farmers, there were other transfers. Black Africans were responsible for the establishment of rice culture in South Carolina, and indigo planting was brought from the European colonies in the West Indies. Peter H. Wood, *Black Majority: Negroes in Colonial South Carolina from 1670 through the Stono Rebellion* (New York: Knopf, 1974), 59–62; John R. Commons et al., eds., *A Documentary History of American Industrial Society* (Cleveland: Clark, 1910), 1: 265–66.

drove some of them from their trades into farming.[11] The crafts that flourished initially were those associated with agriculture and other exploitations of natural resources, and construction.

Milling, for example, was a crucial technology for the early settlers, providing for both culinary and construction needs, and processing agricultural and forestry products for export. There is little record of the millers' migration, but it is abundantly clear that their skills were highly valued. In the seventeenth and eighteenth centuries towns and even state governments established legal advantages and monetary incentives to entice millers.[12]

The colonial millers very early turned their ingenuity to water-powered sawmilling, a task normally accomplished by muscle power in Europe. New England sawmills and succeeding generations of millers developed new technologies such as gang and muley saws.[13] Late in the eighteenth century the Delaware miller Oliver Evans introduced continuous-flow processing into gristmilling.[14] Even with these American modifications, however, an immigrant English millwright such as Thomas Oakes (who had reputedly worked under the famous English engineer John Smeaton) was much in demand in the first two decades of the nineteenth century.[15]

Other important immigrant artisans of the predominantly agricultural economy of the colonial era included blacksmiths, carpenters, masons and bricklayers, shipwrights, tanners, and shoemakers.[16] With the development of a small urban elite by

[11] Carl Bridenbaugh, *The Colonial Craftsman* (Chicago and London: University of Chicago Press, 1961), 3.

[12] Edward P. Hamilton, *The Village Mill in Early New England* (Sturbridge, Mass.: Old Sturbridge Village, 1964), 19; Louis C. Hunter, *Waterpower in the Century of the Steam Engine. A History of Industrial Power in the United States, 1780–1930* (Charlottesville: University of Virginia Press, 1979), 28–33.

[13] Nathan Rosenberg, *Perspectives on Technology* (Cambridge: Cambridge University Press, 1976), 36–37; Brooke Hindle, "The Artisan During America's Wooden Age," *Technology in America,* 12; Hamilton, *The Village Mill in Early New England,* 15.

[14] The most recent and most thoughtful work on Evans is Eugene S. Ferguson, *Oliver Evans: Inventive Genius of the American Industrial Revolution* (Greenville, Del.: The Hagley Museum, 1980).

[15] Thomas Gilpin, "Fairmount Dam and Water Works, Philadelphia," *Pennsylvania Magazine of History and Biography* 37 (1913): 471–77; *An Additional Report on Water Power, by the Watering Committee . . .* (Philadelphia: n.p., 1819), 3.

[16] Bridenbaugh, *The Colonial Craftsman,* 3, 6, 7; Carroll W. Pursell, Jr., ed., *Readings in Technology and American Life* (New York: Oxford University Press, 1969), 8–9; Carroll, *The Timber Economy of New England,* 63–64, 69. The original expedition

1700 more specialized craftsmen, such as cabinetmakers, silversmiths, buttonmakers, and pewterers, became established in the seaboard cities.[17] Later, in mid-century, a few communities, notably the Moravian towns of Bethlehem, Pennsylvania, and Salem, North Carolina, became known as settlements of a variety of skilled immigrants. Other towns specialized in a particular craft: Lynn, Massachusetts, was early known for its shoemakers.[18]

With these technologies, which were either fundamentally agricultural or oriented toward the wealthy, the American seaboard had by 1775 what might have been considered by the casual mercantile theorist the ideal colonial economy: one which was heavily extractive, requiring an active trade with more industrialized areas to provide manufactured goods, yet possessing population centers of sufficient refinement to provide a comfortable environment for a bureaucrat sent to a provincial outpost. What such a casual observer would have missed, however, was the presence of numerous industrial enterprises, collectively small compared to the agricultural economy, but clearly, in historical perspective, the roots of industrialization.

Germination and Growth: The Transfer of Iron and Glass Technologies to Colonial America

Seventeenth- and eighteenth-century Europe had an active iron industry, certainly the most active in the world, and one which was on the verge of rapid development and growth. Britain had many ironworks, but it was experiencing a growth in demand for refined (bar) iron which could not be satisfied.[19]

which landed at Jamestown in 1607 included four carpenters, one blacksmith, two bricklayers, and one mason. The next year several more blacksmiths arrived. These craftsmen seem, however, to have been sent to aid in the expected mining activity, rather than agriculture. Sigmund Diamond, ed., *The Creation of Society in the New World*, The Berkeley Series in American History (Chicago: Rand McNally, 1963), 7–8.

[17] Bridenbaugh, *The Colonial Craftsman*, 6–7, 67, 87; James A. Mulholland, *A History of Metals in Colonial America* (University: University of Alabama Press, 1981), 88, 96.

[18] Bridenbaugh, *The Colonial Craftsman*, 26–29, 49, 54–56.

[19] Charles K. Hyde, *Technological Change in the British Iron Industry, 1700–1870* (Princeton: Princeton University Press, 1977), 19–20.

It is not surprising, then, that early settlers made attempts to establish iron manufacture in America as soon as the most superficial mineralogical surveys of the land revealed substantial quantities of iron ore and limestone, the two raw materials for smelting. For fuel there was no question about the ready availability of wood for charcoal.

The first effort to make iron was in Virginia. In 1620–21 the Virginia Company sent one hundred fifty workmen, "all purportedly experienced," to establish a works. A site was selected on Falling Creek (southwest of modern Richmond), and early in 1622 the works were on the verge of production. An Indian attack then destroyed the establishment, massacred the workmen, and ended for nearly a century any ironmaking activity in Virginia.

Twenty years later John Winthrop, Jr., brought to New England a group of skilled ironworkers, with tools and materials, having previously established a company in England to support the venture. He founded the Braintree, Massachusetts, furnace and forge in 1644, and the Hammersmith works at Saugus in 1645. The latter establishment had a furnace, forge, and (later) a rolling and slitting mill. It appears to have been one of the few ironworks of its era which had production and manufacturing facilities at one site.

The Braintree furnace was abandoned in 1647, and the Hammersmith works were closed by the 1670s, but unlike in Virginia, both the interest and skills remained. Iron manufacture spread from the nurturing sites in Massachusetts to others throughout New England.[20] Other seventeenth-century works were subsequently established in New Jersey.[21]

The eventual heart of colonial iron manufacture, however, was in Pennsylvania, and to a lesser extent in Maryland and Virginia. Pennsylvania benefitted from the immigration of British and Welsh Quaker ironmasters, as well as a mixture of skilled Germans, Irish, and French. They found the abundant natural resources and waterpower (for operating the

[20] Mulholland, *A History of Metals in Colonial America,* 23–24, 31–32, 35, 58, 69; W. David Lewis, *Iron and Steel in America* (Greenville, Del.: The Hagley Museum, 1976), 17–18.

[21] Lewis, *Iron and Steel in America,* 18; Mulholland, *A History of Metals in Colonial America,* 66–67, 69.

blast and forge machinery) of the Great Valley and Welsh Mountains of southeastern Pennsylvania to be ideal for iron manufacture. The first forge (1716) and furnace (ca. 1720) were established by Thomas Rutter in Berks County, and the founding of other works followed rapidly. By the Revolution over twenty furnaces, forty-five forges, and various other associated enterprises had been built in the colony.[22]

The great Principio works were established in Maryland, at the head of the Chesapeake Bay, in 1718–21 by an English company. Ironmasters Stephen Onion and John England were successively in charge, and under England operations were put on a firm footing. Another major works was established in 1733–34 by the Baltimore Company, a group of three Maryland merchants. They engaged the services of William Taylor, an English ironmaster, to put in operation a furnace on Gwynn's Falls. He insured the success of the effort by importing eighty-five tons of English stone for the furnace's hearth and inwalls.[23]

The Virginia industry was reinstituted by Colonel Alexander Spotswood, who erected three furnaces along the Rappahannock River, beginning in 1719. He brought in Germans to direct the works, and they settled in a town named Germanna. Other works in Virginia were set up by the Principio Company, and before the Revolution Pennsylvania German migrants built ironworks in the Shenandoah Valley.[24]

Glassmaking was another technology which required facilities on an industrial scale. In addition to the raw materials (sand, lime, and potash) a glassmaker needed a furnace capable of withstanding the heat needed to fuse glass, ceramic pots to hold the melted glass, and a variety of implements for blowing, handling, and shaping the hot semi-liquid. This normally required the investment of considerable capital, a period of several months for construction and trial runs, and the pur-

[22] Mulholland, *A History of Metals in Colonial America,* 66–67, 69; Arthur Cecil Bining, *Pennsylvania Iron Manufacture in the Eighteenth Century,* Publications of the Pennsylvania Historical Commission, vol. 4 (Harrisburg: Pennsylvania Historical Commission, 1938), 49–51, 187–89.

[23] Mulholland, *A History of Metals in Colonial America,* 62–63; Keach Johnson, "The Genesis of the Baltimore Ironworks," *Journal of Southern History* 19 (May 1953): 162, 168, 170, 172–74.

[24] Mulholland, *A History of Metals in Colonial America,* 63–66; Lewis, *Iron and Steel in America,* 21–22.

chase or leasing of acres of forest to provide sufficient fuel. In these terms it was not unlike establishing an ironworks.

Seventeenth-century attempts to establish glassmaking in the American colonies were limited. Two small glassworks, dependent on the skills of Dutch, Polish, and Italian craftsmen, were set up at Jamestown in 1608 and again in 1621–24. An English glassmaker helped establish a glasshouse near Salem, Massachusetts, in 1639, but it survived only two or three years. Other seventeenth-century glasshouses of undetermined lifetime were in New York and Philadelphia.

A continuous tradition of American glassmaking began through the efforts of Caspar Wistar, a German with entrepreneurial talents who arrived in Philadelphia in 1717. He immediately established a brass-button manufactory and rapidly accumulated wealth. For reasons which are unknown, Wistar decided to enter the glass business and in 1738 contracted with four German glassmakers to set up a works. At least one additional group of Germans later joined the venture, and numerous assistants and apprentices were also hired.

Wistar erected his glasshouse, known as Wistarburgh, near Salem, New Jersey, along a tributary creek of the Delaware Bay. The location had sufficient sand deposits, enough wood to fuel the furnace and make into potash, and lime in the form of oystershells, limestone, and chalk. The workers erected a typical northern European-style furnace about twelve feet long, eight feet wide, and six feet high. After some years of indifferent success they located suitable refractory clay for the pots in which the glass was made.

Wistarburgh produced a variety of glassware, including bottles, window glass, and even (at the instance of Benjamin Franklin) tubes for generating static electricity. The glasshouse operated continually until 1776 or 1777 when the Revolutionary War disrupted its activity. Although it never reopened, several of the skilled workers subsequently established other southern New Jersey glasshouses, sustaining a tradition of glassmaking in the area which continued into the twentieth century.

More famous than Wistarburgh, but shorter-lived, were the glasshouses of Henry William Stiegel. He came to Pennsylvania from Germany in 1750 and two years later took a position as a bookkeeper at the Elizabeth (iron) Furnace in Lancaster County. Having accumulated some capital, in 1758 he joined with three others to purchase the Elizabeth Furnace and thereby embarked on a career of real estate transactions and industrial promotion which ended only with his bankruptcy in 1774.

Stiegel erected a small glasshouse at Elizabeth Furnace in 1763, employing several skilled workers, most of them apparently German immigrants. The next year he opened a second glasshouse at nearby Manheim, a town which he had just helped to found. Both glasshouses were probably profitable, but they were rather small operations and made mostly bottles and window glass. They did not match the vision of the great entrepreneur.

In 1769 Stiegel began building a larger glasshouse at Manheim, and brought in more skilled European workmen (bringing his total glassmaking workforce to over a hundred), including English, Irish, Germans, and Italians. When the works went into production they made a variety of common and fine tableware. It was the production of this glasshouse over the next five years which made Stiegel glass famous. The craftsmen not only blew beautiful shapes, but were masters at enamelling, the use of pattern molds, and engraving.

With the general collapse of Stiegel's enterprises in 1774 the glasshouse soon ceased production, but a number of his workmen played significant roles in other glassworks. At least one had already gone to work at Philadelphia, and others followed him. One worker who probably had previous experience at Wistarburgh returned to southern New Jersey to start a works at what became Glassboro. Others eventually joined the vast glass enterprise begun in 1784 in western Maryland by Johann Friedrich Amelung, a German immigrant. Some of those workers even experienced a third migration from Maryland to western Pennsylvania in the late 1790s, setting up at New Geneva the first glasshouse in the upper Ohio Valley.[25]

[25] J. L. Harrington, *Glassmaking at Jamestown: America's First Industry* (Richmond, Va.: Dietz, 1952), 5–10; Lura Woodside Watkins, *American Glass and Glassmaking* (New York: Chanticleer, 1950), 21–34; Arlene Palmer, "Glass Production in Eighteenth-century America: The Wistarburgh Enterprise," *Winterthur Portfolio* 11 (1976): 75–101; George L. Heiges, *Henry William Stiegel and His Associates: A Story of Early American Industry* (Lancaster, Penna.: n.p., 1948), 8–9, 12, 30, 58–59, 91, 96–101, 123, 134, 143, 148, 188–93.

Grafts and New Shoots: Borrowed and Indigenous Industrial Development, 1775–1825

In the last quarter of the eighteenth century there was evident in the United States (no longer colonies) a vitality of technical endeavor that created opportunity both for native artisans of inventive bent, and for immigrants with knowledge of the latest devices and processes in Europe. Certainly there were, up to about 1825, few Americans like Jacob Perkins, whose head was so full of ideas that they could find sufficient expression only in the most advanced industrial nation of the world, Great Britain. For most American technologists the field for innovation was so great that they willingly suffered with the problems of the nation's technical adolescence. Eli Whitney may have found the promotion and protection of his cotton gin patent a wearying task, but there was probably nowhere else that his device would have had significant economic use.

Oliver Evans is one of the earliest examples of American mechanical genius. Born in 1755 and raised in northern Delaware, he was apprenticed to a wagonmaker and wheelwright. He turned to gristmilling at age twenty-seven. The gristmills he knew were primarily merchant mills, grinding grain on a large scale for packing and shipping, and Evans became fascinated with the possibility of rationalizing the multi-stage milling process. Applying the excess power of the waterwheel which drove the millstones, Evans perfected a system of elevators, sweeps, hoists, and other devices that made the milling operation largely automatic. In the latter 1780s his devices were installed in some mills, and in 1790 he received a patent for his improvements. Subsequent adoption of his devices was stimulated by the publication of his *Young Mill-wright and Miller's Guide* (1795), which discussed the theory and practice of milling, as well as the operation and virtues of his system.

Evans by the time of publication had established a business in Philadelphia, where he came in contact with a community of craftsmen, particularly metalworkers, and he turned his attention to steam engines. In 1803 he publicly displayed a high-pressure engine which he used to drive a screw mill for pulverizing plaster of Paris. The engine had been built with the assistance of several artisans, including Charles Taylor, an English steam engineer, and Evans subsequently put Taylor in charge of a foundry at which he proposed to accept orders for engines. Ultimately Evans established the Mars Works, which became the leading foundry and machine shop in Philadelphia, and until his death in 1819 he was largely engaged in manufacturing engines and other metalwork.[26]

Evans's contributions to American technology were substantial. His automatic milling devices became standard here, and were the object of investigation and adoption by Europeans. They are an early example of the distinct American genius for developing continuous flow technology.[27] High-pressure engines became important for water and rail transport, especially western steamboats, which are another peculiarly American innovation.[28]

Thus in milling, where an American pattern of innovation had been visible since early in the colonial era, Evans proceeded without European inspiration. But when he moved into the realm of modern industrial technology, he had to rely upon Charles Taylor, a skilled English immigrant. Other Americans also found that native ingenuity could not substitute for intimate knowledge of industrial technologies when they wished to establish them in America. One of the most striking individual instances of a personal transfer of European technologies was the visit of the British civil engineer William Weston to the new nation in the 1790s (see chapter 2).

Factory textile technology was another area in which the source of innovations was almost totally European. Widespread home manufacture of textiles with hand spinning wheels and looms only partially prepared Americans to replicate the textile factories springing up in Great Britain in the late eighteenth and early nineteenth centuries. In fact, the foremost historian of the transfer of textile technology has concluded

[26] Ferguson, *Oliver Evans,* 11–51.

[27] Ibid., 63; W. O. Henderson, "Peter Beuth and the Rise of Prussian Industry, 1810–1845," in *The Development of Western Technology Since 1500,* ed. Thomas Parke Hughes (New York: Macmillan, 1964), 115.

[28] Eugene S. Ferguson, "The Steam Engine before 1830," in *Technology in Western Civilization,* eds. Melvin Kranzberg and Carroll W. Pursell, 2 vols. (New York: Oxford University Press, 1967), 1: 258–61.

that the crucial element in the process was the emigration of textile machine builders. Americans could learn to operate the new machines and possessed the requisite mechanical skills to build them, but without the largely nonverbal knowledge of the best and most appropriate designs, the information carried by the immigrants, there would have been no imitation of the English textile revolution.[29]

Samuel Slater's transfer of cotton textile technology to the United States has long been a part of the standard chronology of American industrialization. Slater emigrated in 1789, having experience with cotton processing machinery in the pioneer British mill of Jedediah Strutt. He settled in Rhode Island and constructed from memory the cotton spinning machinery for a factory in Pawtucket. Subsequently he became a partner in a firm which erected mills at Pawtucket and other sites. He was a competent factory manager and his mills were early models for other textile entrepreneurs.[30]

Yet if Slater was one of the earliest, he was certainly not the only British immigrant with important cotton textile machine skills. Some had come as early as before the Revolution, but with insufficient financial encouragement, their efforts to build and sell machines were invariably unsuccessful. Slater's emigration coincided with the rise of a better environment for textile innovation. William Pearce arrived in 1791 and built machines at Paterson, New Jersey, Philadelphia, and Wilmington, Delaware, over the next several years. A cotton mill near New York in 1794 was reported to have a dozen or more workmen from Manchester, and the machinery was made "on the spot from models brought from England and Scotland." James Davenport built spinning mules and waterframes

[29] David J. Jeremy, "British Textile Technology Transmission to the United States: The Philadelphia Region Experience, 1770–1820," *Business History Review* 47 (Spring 1973): 36–42; David Jeremy, *Transatlantic Industrial Revolution: The Diffusion of Textile Technologies Between Britain and America, 1790–1830's* (Cambridge, Mass.: M.I.T. Press, 1981), 20–35, 74.

[30] David J. Jeremy, "Innovation in American Textile Technology during the early 19th Century," *Technology and Culture* 14 (January 1973): 41–42, 44–45; Barbara M. Tucker, "The Merchant, the Manufacturer, and the Factory Manager: The Case of Samuel Slater," *Business History Review* 55 (Autumn 1981): 297–313. Tucker's *Samuel Slater and the Origins of the American Textile Industry, 1790–1860* (Ithaca and London: Cornell University Press, 1984) is not essentially concerned with technology, and does not extend our knowledge of Slater's role in transferring textile technology.

(power looms) in Philadelphia in 1796. This era of the individual cotton machine builder reigned for about twenty years until centers of textile machine manufacture such as Wilmington, Delaware, grew up along the coastal areas of the middle and northern states. Immigrant machine builders still brought the latest in cotton textile machines to the United States, however.[31]

The other major branch of the textile industry, woolens, mechanized more slowly than cotton, but followed a similar pattern of transfer. Carding was a crucial stage in manufacturing. The kinky woolen fibers were worked between opposing rows of metal hooks set in stiff leather cards so that the fibers were straightened and made parallel to one another. The wool was then ready for spinning. The mechanization of this process occurred in Britain in the late eighteenth century, and the emigration of the Schofield family in 1793 from Yorkshire brought knowledge of it to the United States.

The Schofields settled in Newburyport, Massachusetts, and established a small woolen cloth factory. Local entrepreneurs quickly recognized their abilities and hired them to construct the equipment for a much larger establishment. Thereafter the family prospered, largely by establishing or furnishing new mills, and gradually its members spread throughout New England. They trained a second generation of woolen machine builders, including Paul Moody, who adapted his knowledge to help establish the pioneering cotton mill at Waltham, Massachusetts.[32]

Other skilled woolen machine builders and operators came in the decade after the Schofields. (A few had come earlier with little effect.) Some worked in conjunction with the Schofields, but others (coming largely to New England as well) found independent success in machine manufacture. From

[31] Carroll W. Pursell, Jr., "Thomas Digges and William Pearce: An Example of the Transit of Technology," *William and Mary Quarterly,* 3rd ser., 21 (October 1964): 551–60; Pursell, ed., *Readings in Technology and American Life,* 29; Herbert Heaton, "The Industrial Immigrant in the United States, 1783–1812," *Proceedings of the American Philosophical Society* 95 (October 1951): 520; Monte A. Calvert, *The Mechanical Engineer in America, 1830–1910* (Baltimore: Johns Hopkins Press, 1967), 3–4.

[32] Jeremy, *Transatlantic Industrial Revolution,* 97, 118–28; Jeremy, "Innovation in American Textile Technology during the early 19th Century," 45.

1810 to 1820 the American woolen industry expanded tenfold, attracting more English machine builders. But skilled operatives arrived in increasing numbers, too, and even continental European textile workers found ready employment.[33]

By the latter 1810s woolen factories had spread out of the New England and Middle Atlantic states and had begun to rely in part on Americans and American improvements for their machinery. A major instance was the Steubenville (Ohio) Woolen Mill. It was the brainchild of Bezaleel Wells, a land speculator and entrepreneur of eastern Ohio who also became interested in sheep raising. By early 1814 Wells had formed a company with Samuel Patterson of Steubenville and James Ross and Henry Baldwin, two Pittsburgh politicians and businessmen. During the next few months Wells hired Christopher H. Orth to superintend the mill. Orth, a German, apparently had some background in woolen manufacture because in the fall of 1814 he patented a shearing device (to cut the raised nap of the woolen cloth). The partners also hired Benjamin Henry Latrobe (a consulting engineer, born and trained in England, then living in Pittsburgh) to direct the manufacture and erection of a steam engine to power the mill.[34]

The crucial knowledge of wool carding machinery, and probably other devices, was apparently obtained by bringing Paul Moody to the mill. William Price, an English craftsman from Stourbridge known in Pittsburgh for his mechanical genius, built some of the machinery. The mill was in full operation by the spring of 1816 and employed Scottish weavers as well as what was advertized as "an experienced workman" at the carding machines. From the beginning the Steubenville Woolen Mill was successful, and by 1825 a national journal referred to it as "the celebrated establishment of Messrs. B. Wells and company, who send to the Atlantic states many thousand dollars' worth of superior cloths every year and a large amount of other woolen goods."[35]

[33] Jeremy, *Transatlantic Industrial Revolution,* 119–21, 128–31, 231, 234–235, 276, 280–83.

[34] Darwin H. Stapleton, ed., *The Engineering Drawings of Benjamin Henry Latrobe* (New Haven: Yale University Press, 1980), 55–57.

[35] Benjamin Henry Latrobe to Bezaleel Wells, 15 October 1814, 120/C12, Latrobe to C. H. Orth, 27 March 1815, 115/D14, Latrobe to C. H. Orth, 5 April 1815, 125/G4, *The Microfiche Edition of the Papers of Benjamin Henry Latrobe,* Thomas

By 1825 the status of American woolen and cotton manufacturing was such that the American textile manufacturer Zachariah Allen could visit Britain and find that American mills had machinery equal to or in advance of most of what he saw abroad. Moreover, American textile technology was already exhibiting distinct differences from counterparts in Britain. This distinctiveness was such that J. C. Dyer, a Rhode Island merchant and textile mechanic who had settled in England in 1811, could set up a textile factory in Manchester in the 1820s which used American-style machines. In the 1830s he established his sons as textile machine makers and cotton spinners in France.[36]

Thus, instead of a unidirectional transfer from the eastern rim of the Atlantic to the west, what was emerging as a pattern was what Anthony Wallace has styled the "international fraternity of mechanicians." These men, American and European, kept in constant communication with the centers of innovation, mostly through travel or informants (such as immigrants), but also through letters and (more frequently after 1830) through publications.[37]

Fully Grown: American Technological Maturity, 1825–50

The rough equality of American textile technology with that of Britain by 1825 was duplicated in many other fields of American industry during the next twenty-five years, and by mid-century (or 1851 if the Crystal Palace Exhibition of that year is used as a benchmark) the United States was indisputably a modern industrial power. Yet from 1825 to 1850 the transfer of technology was just as important to American

E. Jeffrey, ed. (Clifton, N.J.: White, 1976); Lowell Innes, *Pittsburgh Glass, 1797–1891* (Boston: Houghton Mifflin, 1976), 17–19; Lowell Innes, "William Price and the Round Church," *Western Pennsylvania Historical Magazine* 47 (October 1964): 17–18; J. A. Caldwell, *History of Belmont and Jefferson Counties, Ohio* (Wheeling, W.Va.: Historical Publishing Co., 1880), 493; Steubenville (Ohio) *Western Herald,* 12 April 1816; *Niles' Weekly Register* (Baltimore) 28 (1825): 82n.

[36] Jeremy, *Transatlantic Industrial Revolution,* 131–37; Jeremy, "Innovations in American Textile Technology During the 19th Century," 40–76; Mira Wilkins, *The Emergence of Multinational Enterprise: American Business from the Colonial Era to 1914* (Cambridge, Mass.: Harvard University Press, 1970), 20n.

[37] Anthony F. C. Wallace, *Rockdale: The Growth of an American Village in the Early Industrial Revolution* (New York: Knopf, 1978), 211–19.

industry as before. Moreover, it occurred with greater frequency and more systematically, largely because constantly improving transatlantic communication and transportation had two effects: (1) news of European innovations or of American opportunities for skilled Europeans moved more quickly, and (2) both European emigration and the European journeys of skilled Americans could take place more expeditiously.[38] (One result was that the annual number of immigrants, still largely from northern Europe, increased twenty-fold from 1825 to 1850.) Notices of European developments also spread through the establishment of technically oriented journals, notably the *Journal of the Franklin Institute* (1826) and the *American Railroad Journal* (1831).[39] Under these conditions the United States was ripe for the introduction of new industry, but old industries were reformed as well.

Mining, for example, was changed from an activity largely auxiliary to metalworking, with some open-pit working of coal, to an intensive shaft-mining industry. Prior to the 1820s most iron furnaces depended upon local supplies of ore and limestone which had been located as surface outcroppings. Less than a dozen men would be employed at each works to dig and cart the ore no more than a few miles by wagon to the furnace. Even the Cornwall mines in Lebanon County, Pennsylvania, opened by Cornish miners in the mid-1700s, were no more than forty feet deep in the 1780s. In 1796 a principal mine in the most advanced coal mining in the United States, the Chesterfield region along the James River, was described as an open pit "about 50 Yards square and about 30 feet deep. Many Works or Drifts are from thence carried into the body of the Coal, 5 feet wide." In 1818 a mine in the same region was worked under the direction of two Scottish miners. They had dug several shafts, the deepest going to three hundred fifty feet and requiring a steam engine to pump out water. The miners used only picks to open horizontal galleries and the coal was raised by mule power.[40]

[38] George Rogers Taylor, *The Transportation Revolution, 1815–1860* (New York: Harper, 1968), 104–22, 144–48.

[39] Bruce Sinclair, *Philadelphia's Philosopher Mechanics: A History of the Franklin Institute, 1824–1865* (Baltimore and London: Johns Hopkins University Press, 1974), 195–216; John F. Stover, *American Railroads* (Chicago and London: University of Chicago Press, 1961), 21.

[40] Bining, *Pennsylvania Iron Manufacture in the Eighteenth Century*, 69–71; Joseph

The rudimentary state of American mining was a matter of concern to William H. Keating, a Philadelphia investor interested in the development of the anthracite coal regions of eastern Pennsylvania. Upon his return from a European trip in 1821 he wrote a pamphlet urging Americans to adopt the modern mining technology that he had observed. The use of black powder blasting for excavation, of steam engines for pumping and raising the coal or ore, and of railways for transporting it out of the mine was rare in the United States. Keating's wishes were soon fulfilled by a flood of talented English, Welsh, Cornish, and Scottish miners. The first arrived in the anthracite regions in 1827, where in 1833 an Englishman superintended the first use of steam power. Others arrived later in the bituminous regions of western Pennsylvania and the Midwest, and in the later 1840s Cornish miners helped open the copper and iron mines of upper Michigan.[41]

Iron manufacture was closely related to mining. Here also various British immigrants played important roles. American charcoal iron technology had been on par with its British counterpart since the colonial era, but the newest form of pig iron production—that with coal fuel—was not readily adopted in America. The combined knowledge of hot-blast ovens, higher furnace stacks, increased blast pressure, and the nonverbal skill of coal-firing never crystallized in various American experiments, and the beginning of the new era occurred only when a Welshman, David Thomas, built and put in blast an anthracite coal furnace at Catasauqua, Pennsylvania, in 1840 (see chapter 5).[42]

Similarly, many aspects of ironworking and fabricating tech-

E. Walker, *Hopewell Village: The Dynamics of a Nineteenth Century Iron-Making Community* (Philadelphia: University of Pennsylvania Press, 1966), 249–50; Benjamin Henry Latrobe, *The Virginia Journals of Benjamin Henry Latrobe,* 2 vols. (New Haven: Yale University Press, 1977), 1: (quote on p. 97) 96–97; John Drammer, Jr., "Account of the Coal Mines in the Vicinity of Richmond, Virginia," *American Journal of Science* 1 (1819): 125–28.

[41] William H. Keating, *Considerations Upon the Art of Mining* (Philadelphia: M. Carey and Sons, 1821); Rowland Tappan Berthoff, *British Immigrants in Industrial America, 1790–1950* (Cambridge, Mass.: Harvard University Press, 1953), 48–50, 60; Frederick Moore Binder, *Coal Age Empire: Pennsylvania Coal and Its Utilization to 1860* (Harrisburg: Pennsylvania Historical and Museum Commission, 1974), 57.

[42] Peter Temin, *Iron and Steel in Nineteenth-Century America: An Economic Inquiry* (Cambridge, Mass.: M.I.T. Press, 1964), 14–15, 57–62, 96–98; see chapter 5, below, on David Thomas and the introduction of anthracite fuel.

nologies in the United States were introduced or improved by British immigrants. They introduced both new techniques for wrought iron manufacture (puddling in 1817 and boiling in 1837, both at Pittsburgh); superintended the erection of rail mills in the 1840s; and produced the first wrought iron tube. Skilled Welsh ironworkers were particularly in demand. John Fritz, a great American innovator in iron and steel technology in later years, recalled in his autobiography how stimulating it was to work under a Welsh foreman during his apprenticeship at the Norristown (Pennsylvania) Iron Works in the early 1840s.[43]

The career of Henry Burden provides one of the most detailed records of the transfer of ironworking technology. Born in Scotland in 1791, he was educated at the University of Edinburgh and trained in the iron industry. In 1819 he emigrated to the United States and established himself in the developing iron district of Troy, New York. He soon developed an improved plow, an improved cultivator, and a "hemp and flax machine." He became superintendent of the Troy Iron and Nail Factory in 1822, and in 1825 became a consultant for the Springfield (Massachusetts) Arsenal, where he constructed a rolling and slitting mill. In the 1830s he invented machines for manufacturing large quantities of horseshoes and railroad spikes, and the "Burden rotary concentric squeezer" which worked puddled iron prior to rolling it into bars. These latter devices came into worldwide use.[44]

Burden returned to Britain twice during his career (1827–28 and 1835–36) in order to keep abreast of improvements in his field. In so doing, he was one of the earliest visitors from America in his field, and certainly one of the best informed. But he was not alone. Samuel V. Merrick's trip to England in 1834 eventually led to his purchase of the rights to the Nasmyth steam hammer, which he used in one of the

[43] Berthoff, *British Immigrants in Industrial America,* 62–63; [John Fritz], *The Autobiography of John Fritz* (New York: Wiley, 1912), 85–88. As late as 1868 a Welsh foreman was brought to America by a major Colorado corporation to solve its blast furnace problems: Carroll Pursell, "Science and Industry," in *Nineteenth Century American Science: A Reappraisal,* ed. George H. Daniels (Evanston, Ill.: Northwestern University Press, 1972), 243.

[44] Paul J. Uselding, "Henry Burden and the Question of Anglo-American Technological Transfer in the Nineteenth Century," *Journal of Economic History* 30 (1970): 312–37.

major foundries in Philadelphia.[45] Other metallurgically inclined Americans who went to Europe about this time included the mechanical engineer George Escol Sellers, and James Bogardus, who introduced new iron-building technologies to America. A generation later Alexander Holley began his regular transatlantic voyages, first bringing the Bessemer steel technology to the United States, then aiding in the introduction of American innovations into European steelworks.[46]

The transfer of technologies by American visitors to Europe was even more important in mechanical and civil engineering than in metallurgy in the 1825–50 period. In the 1830s particularly there was a tide of Americans with various levels of engineering skills washing onto European shores, especially Britain's.[47]

Although preceded by Loammi Baldwin, Jr. (twice), Sylvanus Thayer, and Canvass White, William Strickland's visit to England in 1824–25 under the auspices of the Pennsylvania Society for the Promotion of Internal Improvement was widely publicized and demonstrated the usefulness of direct study of European innovations. Strickland's ensuing publication of his reports and drawings was well received on both sides of the Atlantic.[48]

[45] Ibid., 321, 330; Sinclair, *Philadelphia's Philosopher Mechanics,* 321; J. Leander Bishop, *A History of American Manufactures,* 2 vols. (Philadelphia: Young, 1864) 2: 547–48.

[46] Eugene S. Ferguson, ed., *Early Engineering Reminiscences (1815–1840) of George Escol Sellers,* United States National Museum Bulletin 238 (Washington: Smithsonian, 1965), 108–34; Carl W. Condit, *American Building: Materials and Techniques from the First Colonial Settlements to the Present,* The Chicago History of American Civilization (Chicago: University of Chicago Press, 1968), 81–83; Jean McHugh, *Alexander Holley and the Makers of Steel* (Baltimore and London: Johns Hopkins University Press, 1980).

[47] Darwin H. Stapleton, "The Origin of American Railroad Technology, 1825–1840," *Railroad History* 139 (Autumn 1978): 65–77. Bruce Sinclair points out that peculiar economic conditions in the early and mid-1830s made it relatively easy for Americans to make transatlantic trips. Bruce Sinclair, "Americans Abroad: Science and Cultural Nationalism in the Early Nineteenth Century," in *The Sciences in the American Context: New Perspectives,* ed. Nathan Reingold (Washington, D.C.: Smithsonian Institution, 1979), 36.

[48] Frederick Randall Abbott, "The Role of the Civil Engineer in Internal Improvements: The Contributions of the Two Loammi Baldwins, Father and Son, 1776–1838" (Ph.D. dissertation, Columbia University, 1952), 70–74, 129–33; Stephen E. Ambrose, *Duty, Honor, Country: A History of West Point* (Baltimore: Johns Hopkins Press, 1966), 65–67; Department of War to Thayer, 20 April 1815, reel 2, microfilm 417, National Archives, Washington; "Canvass White," *Dictionary of American Biography*; Robert E. Carlson, "The Pennsylvania Improvement Society

Much of the traveling American engineers' attention was first focused on canals. Canvass White's trip of 1817 was made specifically to gather information on the mechanical and construction details of locks, aqueducts, and other canal appurtenances. The interests of Loammi Baldwin, Jr., were primarily hydraulic, and The Pennsylvania Society for the Promotion of Improvements' instructions to Strickland centered on canals.

Moncure Robinson (the Virginia engineer examined in chapter 4) went to England in 1825 to satisfy his urge to learn about canals, but returned a confirmed advocate of railroads. By the later 1820s, attention had clearly shifted to railroads, particularly after the Baltimore and Ohio Railroad sent a team of its engineers to England in 1828 and some of their letters were published. Throughout the 1830s American engineers went to England to see modern railroads under construction and in operation. These visits affected not only the design of rail and track but also locomotives. Horatio Allen, Ross Winans, George Escol Sellers, Wirt Robinson, and George W. Whistler each brought his English experience to bear on the manufacture of locomotives in America.[49]

Numerous other fields of American engineering were heavily influenced by transfers of European technology in this era. For example, although gasworks had been operating in the United States since 1816, a new era was initiated after Samuel Merrick's visit in 1834 to England, France, and Belgium to see the most advanced installations. George Rumford Baldwin directed the construction of urban waterworks in

and Its Promotion of Canals and Railroads, 1824–1826," *Pennsylvania History* 31 (July 1964): 295–310; William Strickland, *Reports on Canals, Railways and Other Subjects* (Philadelphia: Carey and Lea, 1826).

[49] George Sweet Gibb, *The Saco-Lowell Shops* (Cambridge, Mass.: Harvard University Press, 1950), 84, 93; "Horatio Allen," "Ross Winans," and "George W. Whistler," *Dictionary of American Biography*; "Diary of Horatio Allen: 1828 (England)," *The Railway and Locomotive Historical Society Bulletin* 89 (November 1953): 97–138; Ferguson, ed., *Early Engineering Reminiscences,* 134, 160–66, 172–79; Neal Fitzsimons, ed., *The Reminiscences of John B. Jervis: Engineer of the Old Croton* (Syracuse, N.Y.: Syracuse University Press, 1971), 91–100. Wirt Robinson's career is rather obscure, but see below, chap. 4, for his involvement with ordering and designing locomotives for the Philadelphia and Reading Railroad.

I am aware of only two British engineers whose emigration had a significant impact on the American railroad scene: See "James Pugh Kirkwood," and "James Laurie," *National Cyclopedia of American Biography.*

New England and influenced their design following a European sojourn. Emile Geyelin came to the United States in 1849, bringing with him the rights for manufacture and sale of the Jonval turbine, and significantly furthered the American adoption of turbines for waterpower.[50]

If the heavy industries were the most dramatically affected by transfers of technology in the second quarter of the nineteenth century, others were also altered or transformed by transfers. The textile industry continued to rely upon British immigrants. The factory managers in the Rockdale region of southeastern Pennsylvania were usually skilled Englishmen. The knit stocking industry was established in Philadelphia by Germans, but was put on a firm basis in the 1830s by a wave of knitters from Nottingham and Leicester. A Macclesfield weaver started the first successful American silk mill at Paterson, New Jersey, in 1840, and a community of English, French, and German silkworkers developed around it.[51]

The transfer of calico cylinder-printing technology is a more complex case because it took place over three decades. The change from printing cloth with blocks to using rollers had the potential of allowing much greater production of calicos, as well as reducing costs. An English patent on a workable machine was taken out in 1783; the first calico printer was exported to the United States in 1809 in the care of Joseph Siddall, an American who had gone to England to learn about the process. He also brought over the brothers Issachar and James Thorp, who joined him in establishing a calico printing business near Philadelphia. Siddall and the Thorps carried on their business for years without apparent diffusion of the technology, beginning in 1817, but later English immigrants helped to make Philadelphia a center for calico printing.

A second phase of the transfer took place in New England

[50] S. V. Merrick, "Report Upon an Examination of Some of the Gas Manufactories in *Great Britain, France,* and *Belgium,*" in *Reports of the Trustees of the Philadelphia Gas Works to the Select and Common Councils of the City of Philadelphia* (Philadelphia: Philadelphia City Councils, 1838), 59–102; Sinclair, *Philadelphia's Philosopher Mechanics,* 321; Binder, *Coal Age Empire,* 29–33, 36; Christopher Roberts, *The Middlesex Canal, 1793–1860,* Harvard Economic Studies, vol. 61 (Cambridge, Mass.: Harvard University Press, 1938), 199; "George Rumford Baldwin," *National Cyclopedia of American Biography*; Hunter, *Waterpower in the Century of the Steam Engine,* 326–28.

[51] Berthoff, *British Immigrants in Industrial America,* 31, 40–43; Wallace, *Rockdale,* 117–19.

in the 1820s. Three companies participated in it, and each found that relying on chance English immigrants or visits to Philadelphia did not provide them with the best assessment of current British technology. Each company decided that it had to send to Britain one of its directors familiar with textile technology, and in each case the directors' visits were crucial in obtaining skilled workers and parts for calico printing machines. The transfers were successful, and the technology diffused rapidly to other companies.[52]

The origin of the American chemical industry was also largely in this period, and was centered in Philadelphia. John Harrison, who studied in England from 1790 to 1792, returned to Philadelphia to establish the commercial production of sulfuric acid in 1793. Later, in 1814, Harrison took advantage of the skills of Eric Bollman, a German-trained chemist and entrepreneur, to install a platinum still for concentrating acids. Eleuthère Irénée du Pont, a trained industrial chemist from France, set up his gunpowder works in the Philadelphia area in 1802–04 (see chapter 3), but he was relatively isolated from other industrialists.

Pharmaceutical businesses were subsequently established in Philadelphia by partnerships of English and Swiss immigrants (1818), and by a German and a Swiss (1822). The latter enterprise was shortly taken over by George Rosengarten, an American; in 1840 he was joined by N. F. H. Denis, a French-trained chemist. Charles Lennig (who may have been a German immigrant) began large-scale manufacture of sulfuric acid and other chemicals near Philadelphia in 1829. In the late 1840s Lennig joined with George T. Lewis, the proprietor of a Philadelphia white lead (paint) factory, to sponsor the European trip and researches of R. A. Tilghman. He had discovered certain potential improvements in alkali and fat processes, but had to go to England to investigate their commercial possibilities. Tilghman's American and European patents formed the basis of the Pennsylvania Salt Manufacturing

[52] Jeremy, *Transatlantic Industrial Revolution,* 104–14; Theodore Z. Penn, "The Introduction of Calico Cylinder Printing in America: A Case Study in the Transmission of Technology," in *Technological Innovation and the Decorative Arts,* ed. Ian M. G. Quimby and Polly Anne Earl, Winterthur Conference Report 1973 (Charlottesville: University of Virginia Press, 1974), 235–51.

Company, which made alkalis and alum. Its facilities, in being located not at Philadelphia, however, but in Natrona, Pennsylvania, near Pittsburgh, indicated a diffusion of the chemical industry.[53]

The production of American photographic materials, a part of the chemical industry, began after the arrival of one of Louis Daguerre's agents at New York in 1840 to teach the daguerrotype process. The Scovill company of Connecticut learned crucial steps in the manufacture of daguerrotype plates from a customer who had seen them made in Paris, and Rosengarten and Denis, perhaps aided by Denis's recent French experience, began producing the requisite chemicals.[54]

If these industries are representative, and I believe that a multiplication of cases would continue to show that they are, by 1850 the transfer of technology was an expected consequence of the development of virtually any innovation in Europe. Rapid and easy communication not only made possible the ready migration of Europeans, but Americans could easily investigate any novelty, and some technically inclined individuals, such as engineers, found it advisable to make European tours just to follow the scope and direction of developments there.[55]

At the same time, the growth and maturity of American industry made it easier to implant new technologies, since the basic sectors of an industrial society, such as transportation, active machine tool and metallurgical industries, and abun-

[53] Rosengarten Collection, Historical Society of Pennsylvania, Philadelphia; George Thompson Papers, Thompson Collection, Historical Society of Pennsylvania, Philadelphia; Williams Haynes, *Chemical Pioneers: The Founders of the American Chemical Industry* (New York: Van Nostrand, 1939), 108, 112–13, 116, 118; Williams Haynes, ed., *American Chemical Industry,* 6 vols. (New York: 1945–54), 1: 148, 177–78, 180–81, 194–95, 213–14, 6: 329–35; Bishop, *A History of American Manufactures,* 2: 568–71.

[54] Reese V. Jenkins, *Images and Enterprise: Technology and the American Photographic Industry, 1839 to 1925* (Baltimore and London: Johns Hopkins University Press, 1975), 12–16.

[55] A sample of European transfers of technology to America over the next century is provided by: McHugh, *Alexander Holley and the Makers of Steel*; Richard Schallenberg, "The Anomalous Storage Battery: An American Lag in Early Electrical Engineering," *Technology and Culture* 22 (October 1981): 725–52; and Paul Clinton Echols, "The Development of Shell Architecture in the United States, 1932–1962: An Examination of the Transfer of a Structural Idea," in *The Transfer of Ideas: Historical Essays,* ed. C. D. W. Goodwin and I. B. Holley, Jr. (Durham, N.C.: Duke University Press, 1968), 3–42.

dant energy sources (in this case both coal and waterpower) were available.

There is one aspect of this picture of active and growing transfer of technologies which should not be overlooked, however. American industrialization has had an intensely regional impact. A very small percentage of the industrial technologies, particularly after the American Revolution, were initially transferred to the American South. Most of the transferors went to the seaboard areas of the Middle Atlantic and New England states, although as the nineteenth century progressed they took more transfers to the Midwest. This phenomenon may in part be attributable to the general reluctance of antebellum immigrants to move to the South.[56] In any case, the regionalized industrial growth of the United States from the late eighteenth to the mid-nineteenth centuries was to some extent caused by the geographical pattern of the transfer of European technologies.

Conclusions

The role of the transfer of technology in the growth of industrial America as related here may be interpreted as one of continual success. Certainly the transfers that determined the shape and direction of American industry were the successes, but in any given historical period it may not have been clear which attempts at transfer would survive and which would not. The seventeenth and eighteenth centuries, for example, yield a continuous record of attempts to establish a silk industry along the Atlantic coast. These dreams and efforts touched (or even consumed) too many lives to call them inconsequential, but transfers of silk technology were either outright failures or marginal successes which did not have the vitality to create a distinctly American industry.[57] Historians

[56] Glyndon G. Van Deusen, *The Jacksonian Era, 1828–1848* (New York: Harper, 1959), 15–16.

[57] L. P. Brockett, *The Silk Industry in America: A History Prepared for the Centennial Exposition* (New York: Silk Association of America, 1876), 26–37; John C. Van Horne, "Joseph Solomon Ottolenghe (ca. 1711–1775): Catechist to the Negroes, Superintendant of the Silk Culture, and Public Servant in Colonial Georgia," *Proceedings of the American Philosophical Society* 125 (October 1981): 398–409.

understandably concentrate on success stories, but the failures should not be ignored because they often provide perspectives from which the successes may be better understood.

Another possibly dangerous tendency in studying the transfer of a technology is to focus on one particular act of transfer as the *sine qua non* of a particular industry. On the contrary, multiple transfers seem to be the rule. There may in fact be an event which in hindsight appears to have been the turning point in establishing the new technology, but the process is always a complex one and careful examination will show that transfers took place over a period of years, usually through the contributions of several people. The instance of the transfer of calico printing examined above is indicative of that complexity. Moreover, while it is one thing to transfer a new technology, it may be another to establish it as a viable and growing technology. Multiple transfers may be necessary to bring the newly implanted technology up to "running speed" so that it is continually abreast of the latest from the originating country. A technically stagnant transfer may soon be overwhelmed or may at best play little role in a growing economy.

In a like manner, transfers must be continually adapted and modified if they are to fit into the economy and society of the host country. On the one hand, this process may be carried out by numerous sub-inventions and technical alterations of the devices and processes themselves. On the other hand, the adaption may be in terms of the acquisition of new skills and knowledge. In Nathan Rosenberg's words:

> Closely associated with this gradual improvement in the innovation itself is the development of the human skills upon which the use of the new technique depends in order to be effectively exploited. There is, in other words, a learning period the length of which will depend upon many factors, including the complexity of the new techniques, the extent to which they are novel or rely upon skills already available or transferred from other industries, etc.[58]

Thus, our historical sense of the importance of transfers of technology must be tempered by the context that failures may

[58] Rosenberg, *Perspectives on Technology*, 197.

provide, by the recognition that transfers are not "once-and-for-all" events but successive instances, and by the understanding that diffusion and adaptation are necessary if the transfers are not to remain isolated and backward.

With these perspectives in mind, we can focus on the conditions under which industrial technologies were transferred to early America. There are several circumstances common to all the transfers discussed here, and indeed common to the transfer and diffusion process throughout history.

First, it must be recognized that any cross-cultural or international contact involves participation by both cultures or nations. Every group is receptive to change to some degree, even though each group expresses its own identity and (as one anthropologist puts it) "reworking is the rule and reinterpretation [is] inevitable."[59] In the case of technological transfer from western Europe and Britain to the British colonies of North America and then the United States, there was a commonality of culture that minimized the need for immediate adaptions and therefore eased the transfer process in most instances.

The technologies that were successfully transferred were those that had been proved already in the originating country. Attempting to take elsewhere a technology still in the trial stages frequently complicated the transfer process beyond the normal scope of human ingenuity. For example, John Harris has shown how the establishment of French plate-glass technology in eighteenth-century Britain was a long struggle because French glassmakers were simultaneously trying to change their manufacturing process from charcoal fuel to coal fuel.[60] Yet there is a tendency to transfer "hot" technology, the most recent successful innovation, rather than the long-standing, tried-and-true technology.

There are perhaps two reasons for this latter phenomenon. Since the transfer of technology was a process requiring considerable commitment on the part of the transferor, it may be that the enthusiasm evoked by a technical innovation was

[59] Melville J. Herskovits, "The Processes of Cultural Change," in *The Science of Man in the World Crisis,* ed. Ralph Linton (New York: Columbia University Press, 1945), 157.

[60] John R. Harris, "Saint-Gobain and Ravenhead," in *Great Britain and her World,* ed. Ratcliffe, 27–70.

one of the major forces that sustained him through his trials.[61] It also may be that those who understood conditions in the recipient country carefully followed the course of innovation in the originating country and seized the appropriate technology when they saw it come to the fore. This generally describes the transfer of anthracite iron technology to the United States, for example.[62]

Thus the economic vision of the transferors may have been as important as their technical skill. In some instances the transferor was also an entrepreneur, and had a notion of the role of his new product or service in America and had the business acumen to make it a success. In other instances the transferor was not the entrepreneur, but was joined by someone who understood the transferor's technology enough to assess it, and whose contribution was finance, organization, and marketing.

In either case considerable social support was often involved in a transfer. In some cases patronage by the government, local bankers, or certain interested merchants carried a transfer of technology through the risk stage. The particularly American receptivity to technological innovation has often been commented upon, and that may have played a role.[63] Given, however, the sheer volume of transfers throughout the world during any period of time, the "American-ness" of technical receptivity may be exaggerated.[64]

If these common conditions tend to hold for any particular transfer of industrial technologies to early America, it is also possible to see that succeeding time periods had their typical means of transfer. In the American experience there seem to have been three general periods of transfer: the colonial era (1600–1800/1810), the visitor era (1800/1810–50), and the age of maturity (post-1850).

The colonial era of technical transfer was dominated by European immigrants. Western Europe clearly had a more

[61] Here I am borrowing the concept stated by Eugene S. Ferguson in "Toward a Discipline of the History of Technology," *Technology and Culture* 15 (January 1974): 2.

[62] See below, chapter 5.

[63] E.g., Nathan Rosenberg, *Technology and American Economic Growth* (New York: Harper, 1972), 33–34, 43–46; Hugo Meier, "Technology and Democracy, 1800–1860," *Mississippi Valley Historical Review* 43 (March 1957): 622.

[64] Ferguson, "The American-ness of American Technology," 3–24, esp. 7–12.

complex and varied technology than North America, but Britain, France, the German states, and other nations passed along their superior technical knowledge through the migration of their citizens. America had a level of technical activity capable of supporting almost all the basic agriculturally-related craft technologies, but only a few of the industrial technologies.

During the visitor era some citizens of the infant United States acquired a training in basic industrial skills that were common in the advanced industrial sectors of Europe. (For example, a prospective American civil engineer in the 1820s may have become well versed in surveying and acquired some experience in canal construction, but he could learn little about railroad construction in America.) The opportunities to learn basic industrial skills came sometimes from the training of Americans by European immigrants, and sometimes from a developing American technological tradition. Some of these trained Americans traveled to Europe to learn about the most recent developments in their fields and were able to transfer that advanced knowledge to the United States. These American visitors were probably more effective than immigrant transferors because they critically assessed European developments in light of their own experience with American resources and society.

The age of maturity arrived when America had clearly developed a native technical tradition equal in vitality to that of the industrial nations of Europe. This equality was measured in part by the growing recognition that American technology had distinct characteristics not found in European technologies. But it was also measured by the increasing incidence of multilateral transfers of technology. After 1850 it was widely recognized that cherished innovations in any particular industrial field were as likely to occur on one side of the Atlantic as the other. Moreover, in the era of maturity the transfer of technology probably became more frequent.[65]

[65] Of numerous studies on the transfer of early American technology to Europe, these serve as useful introductions: Russell I. Fries, "British Response to the American System: The Case of the Small-Arms Industry after 1850," *Technology and Culture* 16 (July 1975): 377–403; Felix Rivet, "American Technique and Steam Navigation on the Saône and Rhône, 1827–1850," *Journal of Economic History* 26 (1956): 22–23; Jeremy, *Transatlantic Industrial Revolution*, chap. 13. See also Darwin H. Stapleton, "American Technology and the Industrialization of Europe in the 19th Century," read for the Organization of American Historians, April 1981, Detroit.

While not every industry fits this scheme exactly (there was no American tinplate industry of significance before the arrival of Welsh tinplaters in the 1890s, for example), most of the centrally important industrial technologies including metallurgy, machine tools, and civil engineering conform to this pattern and chronology.

The following studies contribute to the understanding of the transfer process during each of these phases. Chapters 2 (William Weston and Benjamin Henry Latrobe) and 3 (Eleuthère Irénée du Pont) represent the colonial era of technology transfer. Chapter 4 (Moncure Robinson) presents one of the more spectacular, if relatively unknown, cases of a technically trained American transferring innovative technology from Europe. Chapter 5 (David Thomas) focuses on an immigrant called to America by a group of investors, many of them technologists, who had carefully watched British iron manufacture for an innovation which could use the coal resources in which they had invested. The studies are a sample of the series of technical transfers which were a primary force in carrying early America into the industrial age.

Part 2

Case Studies

II: William Weston, Benjamin Henry Latrobe, and the Philadelphia Plan for Internal Improvements

In 1817, the engineer Benjamin Henry Latrobe considered the United States's need for technical knowledge, and argued for acquiring more European engineers. He wrote:

> I am well aware that there is a reluctance very natural and patriotic to the employment of foreigners. But we import blankets, scissars, and wine, why should we not import knowledge? A good civil engineer is an acquisition peculiarly desireable. . . .[1]

Latrobe was himself an immigrant civil engineer, having come to this country from England twenty years earlier, and his comment in part reflects his sensitivity to slights he had received for being a foreigner. His statement also demonstrates his understanding that a vast historical process was underway—the transfer of European technology to the New World, a process begun centuries before, and perhaps entering its most energetic phase. Yet in another sense, Latrobe's call for importing civil engineers was outmoded: as he wrote work had begun on the Erie Canal, and it was being directed by engineers who were native Americans.

How had the United States gotten to the point of undertaking perhaps the boldest transportation project in the Western world without utilizing engineers from the Western nations who had experience in such efforts? To answer that question in some measure we can examine a series of events which began fifty years earlier, not in the Mohawk Valley of New York along the future line of the Erie Canal, but in Philadelphia and southeastern Pennsylvania.

Philadelphia in the 1760s had the largest population and greatest commercial activity of the British seaboard colonial cities. It was arguably the most cosmopolitan city as well, with many immigrants and a large group of native American mer-

[1] Benjamin Henry Latrobe to William Seaton, 3 September 1817, Letterbooks, in *The Microfiche Edition of the Papers of Benjamin Henry Latrobe,* ed. Thomas E. Jeffrey (Clifton, N.J.: James T. White and Company, 1976), hereafter *Microfiche Ed.*

chants and professionals who had been abroad, especially to Britain. Many of those who had not been abroad had easy access to European newspapers and books, had correspondents abroad, or regularly met and talked with the ship captains, supercargoes, and others who had crossed the Atlantic.

From these sources some Philadelphians learned of the "peculiarly desireable" abilities of European civil engineers to which Latrobe later referred. By the 1760s the canal age had begun in Britain, an age commonly dated from the completion of the Sankey Navigation in 1757, and the Duke of Bridgewater's Canal in 1761. But there were other canals, particularly in Holland and France, which were also objects of attention for travelers. Moreover, at about the same time British turnpike companies began to develop standard procedures for improving roads, and generally they made their ventures economically as well as technically successful.

Philadelphians' interest in canals and roads, or, as they were known in later years, "internal improvements," crystallized in the later 1760s. A prime reason for their interest was the rapid development of central Pennsylvania, which with the end of the French and Indian War was open to settlement.

Since they already monopolized the trade of heavily settled southeastern Pennsylvania, Philadelphia merchants naturally looked to the whole region as their own to control. But central Pennsylvania's natural trade route, which carried increasing quantities of grain, whiskey, lumber, and iron from the interior, was the Susquehanna River. For eons before the colonies' political boundaries had been drawn the Susquehanna had emptied into the Chesapeake Bay, now within the limits of Maryland, and not into the Delaware estuary controlled by Philadelphia. The Quaker City's merchants had to ponder the possibility that central Pennsylvania's trade would go in a new direction, and that their city would decline relative to a new Chesapeake port such as Baltimore. Adopting a strategy with which we have become familiar in the industrial age, Philadelphians seized upon a technical solution to their perceived economic problem—they came to believe that the new technology of canals and turnpikes could divert the developing interior trade to their city as well as promote its trade. No particular event can be identified as the catalyst for the emergence of this strategy, but a cluster of individuals and activities from 1768 to 1772 brought it to a coherent form.

One leader of the movement was Thomas Gilpin. A Pennsylvania merchant of technical bent, he had been in England in 1753 and had kept a journal in which he commented on, among other things, a steam engine which he saw. On his return he purchased an estate on the Delmarva peninsula between the Delaware and Chesapeake bays below Philadelphia. He soon took an interest in the possibility of connecting the two bays with a canal, and with his neighbors spent some time surveying and making plans for the project. On moving to Philadelphia in 1764, Gilpin became a significant member of the city's intelligentsia, some of whom were by this time interested in transportation improvements.[2]

Attempts had already been made to improve water communications in the Philadelphia region, but without much effect. In 1761 the citizens of Berks County wanted to clear the Schuylkill River, their natural highway to Philadelphia, and collected their own funds to do so. The state assembly gave them permission, and they did remove some rocks at the falls of the Schuylkill above Philadelphia.[3] But their efforts were ineffectual because, as Gilpin remarked later, "in clearing a passage through the rocks at the falls, the river is thereby rendered shallower above the obstructions, and the navigation rather less practicable than before."[4] Clearly, the persons in charge were inexperienced in making rivers navigable.

By the time Gilpin made this observation in 1768 he had apparently made a second trip to Britain which permitted him not only to be judgmental, but also to suggest a remedy. Writing in the *Philadelphia Chronicle,* a fortnightly newspaper, Gilpin argued for canalizing the Schuylkill by one of the means he had seen abroad:

[2] Thomas Gilpin, Jr., "Memoir of Thomas Gilpin," *Pennsylvania Magazine of History and Biography* 49 (1925): 293–94, 297–300; Ralph D. Gray, "Philadelphia and the Chesapeake and Delaware Canal," *Pennsylvania Magazine of History and Biography* 84 (October 1960): 403.

[3] *Pennsylvania Archives,* 8th ser., 8: 5189, 5223.

[4] Philadelphus to the Printer of the *Pennsylvania Chronicle,* 9–16 May 1768, *Pennsylvania Chronicle.* I identify Gilpin as the author on the basis of Carl and Jessica Bridenbaugh's statement in *Rebels and Gentlemen: Philadelphia in the Age of Franklin* (New York: Reynal and Hitchcock, 1942), 347.

> The only possible method effectually to compass this desirable event, is that which is so frequently practiced in England and elsewhere, viz. the moderating the current, and deepening the water by a number of dams across the river, accommodated with sluices or locks, to give passage up or down to flats or rafts, as often as occasion requires.[5]

By proposing to raise the river's level, Gilpin also raised the level of discussion to that of workable technology.

Gilpin's authoritative article was undoubtedly read and discussed, because at exactly the same time a group of Philadelphians formed a new learned institution, the American Society, which in its prospectus claimed that one of its goals was to find out "the cheapest and best methods of making highways, causeways, and bridges, joining of rivers, and encreasing our inland navigation."[6] After some skirmishing, this group joined with the revived Philosophical Society, which Franklin had founded, and by late winter of 1769 the American Philosophical Society for Promoting Useful Knowledge, successor to both groups, was well established. Gilpin was elected a member, and he gave the society a draft of his favorite project, a canal across the Delmarva Peninsula.[7]

Six weeks later the Society established a committee of nine, as it was officially reported,

> to view the ground, and consider in what manner a water communication might be best opened, between the provinces of Maryland and Pennsylvania; and particularly by what means the large and increasing number of frontier-settlers, especially those on the Susquehannah and its branches, might be enabled to bring their produce to market at the cheapest rate, whether by land or water.[8]

The interest of the city's merchants in this venture was unmistakable: they subscribed £200 for the committee's expenses and appointed four of the committee's nine members.[9]

[5] Philadelphus to Printer of the *Pennsylvania Chronicle.*

[6] *Pennsylvania Chronicle,* 29 February–7 March 1768; *Pennsylvania Gazette,* 17 March 1768.

[7] J. Peter Lesley, comp., "Early Proceedings of the American Philosophical Society," *Proceedings of the American Philosophical Society,* 22, part 3 (1885): 32.

[8] "An Abstract of Sundry Papers and Proposals for Improving the Inland Navigation of Pennsylvania and Maryland," *Transactions of the American Philosophical Society* 1 (1771): 293; Lesley, comp., "Early Proceedings," 32, 34.

[9] Lesley, comp., "Early Proceedings," 34–38.

In June the committee examined Gilpin's proposed route in the central part of Delaware, and reported their results to the society. With economic concerns uppermost in their minds, the society instructed the committee to carry out new surveys of more northerly routes which would make a more direct connection between the Susquehanna and Philadelphia, since diverting trade to their city was "the great object in view."[10]

Returning to their task in winter, when the "waters and marshes were frozen over," the committee not only looked at the Chesapeake-Delaware routes, but also turned their attention to the navigability of the lower Susquehanna and the possibility of a road from the Susquehanna to the Delaware. In their report of February 1770 they stated that they had located two appropriate routes for a Chesapeake and Delaware canal, that there was a good line for a road from Peach Bottom on the Susquehanna to Christiana Bridge in northern Delaware, and that breaking up only one rocky stretch of the lower Susquehanna would render it a useful navigation. The report was abridged and published in the society's *Transactions* of 1771, along with a map of the routes examined.[11]

The society also took an interest in Gilpin's earlier plea for improving the Schuylkill, in which he pointed out that Reading on the Schuylkill was "nearer to a great part of the country beyond the Susquehanna, than Baltimore."[12] The society sent out a committee of three members, including William Smith, provost of the Academy of Philadelphia, to examine what appeared to be the most convenient water route from the Schuylkill to the Susquehanna, along their tributary creeks, the Tulpehocken and the Swatara. The state assembly also authorized a survey by a team which included the talented surveyor and Philadelphian, David Rittenhouse. By 1772 both groups had reported that it was possible to connect the Tulpehocken and Swatara by canal.[13]

[10] "An Abstract of Sundry Papers," 293–300; Lesley, comp., "Early Proceedings," 39–41.

[11] "An Abstract of Sundry Papers," 295–300; Lesley, comp., "Early Proceedings," 46, 48, 50–51, 55.

[12] Philadelphus to Printer of the *Pennsylvania Chronicle.*

[13] [William Smith], *An Historical Account of the Rise, Progress and Present State of the Canal Navigation in Pennsylvania* (Philadelphia: Zachariah Poulson, 1795), 67; Brooke Hindle, *David Rittenhouse* (Princeton: Princeton University Press, 1964), 94–95; Lesley, comp., "Early Proceedings," 65–66.

Now let us review what had transpired over the previous four years. Between 1768 and 1772 a group of Philadelphians, including natural philosophers, merchants, and skilled men, had thought out and examined to the best of their ability three potential routes for connecting their commercial city with the emerging backcountry of the Susquehanna valley. Assuming that the settlers would utilize the river itself to ship their goods, they had considered how to remove the obstructions from the lower river. And assuming that the products reached that point, they had considered a water route across the Delmarva peninsula to bring them to Philadelphia. Since much of the traffic was grain to be milled for overseas shipment, and Philadelphia had by far the most active grain and flour trade of the colonies,[14] Philadelphians had confidence that backwoods produce would follow so circuitous a route.

But a principal liability of this route was that it required, if not the cooperation, then the tacit compliance of Maryland with the plan. Yet Maryland and Pennsylvania lawmakers were not always cooperative.[15] That problem, as well as the possibility that some of the trade brought to the Chesapeake might choose to remain there, might be avoided by the second route, a road from the lowest possible point on the Susquehanna to the nearest harbor in Pennsylvania's domain. But early Americans did not build roads, they merely cleared them, and, although Pennsylvania's Conestoga wagons were already incomparable freighters, any road was certainly going to be impassable in some seasons, and capable of carrying only moderate traffic at other times.[16]

Since water carriage was cheaper than that on land and more freight could be carried on water, an all-Pennsylvania water route was also an appropriate alternative. On exami-

[14] For a graphic image of the grain trade of 1768–72, see Lester J. Cappon et al., eds., *Atlas of Early American History: The Revolutionary Era, 1760–1790* (Princeton: Princeton University Press, 1976), 27.

[15] Sylvester K. Stevens, *Pennsylvania: Birthplace of a Nation* (New York: Random House, 1964), 71; James Weston Livingood, *The Philadelphia-Baltimore Trade Rivalry, 1780–1860* (Harrisburg: Pennsylvania Historical and Museum Commission, 1947), 33.

[16] George Rogers Taylor, *The Transportation Revolution, 1815–1860* (New York: Harper & Row, 1968), 15–17; Roger N. Parks, *Roads and Travel in New England, 1790–1840* (Sturbridge, Mass.: Old Sturbridge Village, 1967), 3–14; George Shumway and Howard C. Frey, *Conestoga Wagon, 1750–1850,* 3rd ed. (York, Penna.: George Shumway, 1968), 14–25, 61–64.

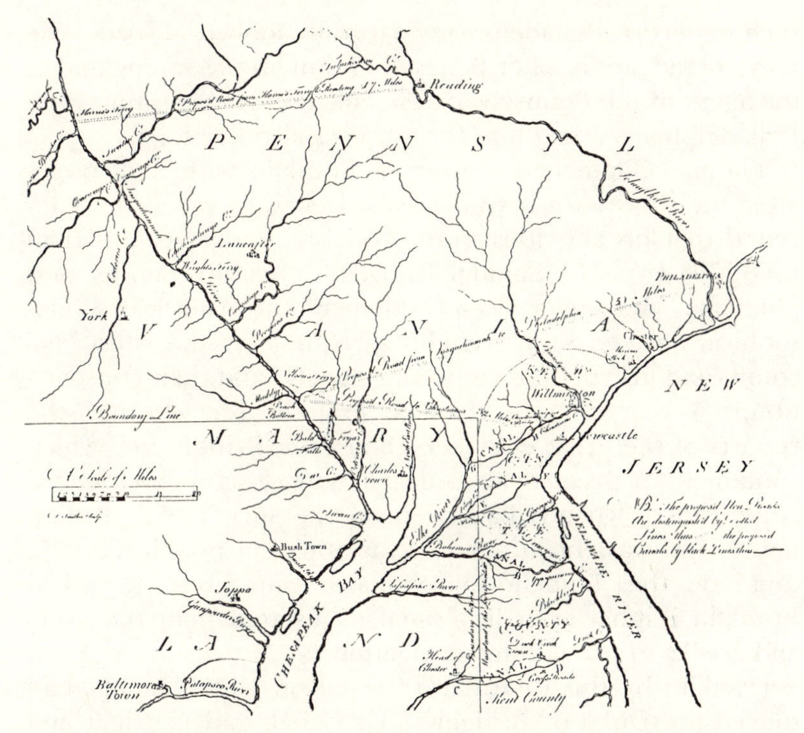

Fig. 1. Map of possible road and canal routes examined by a committee of the American Philosophical Society, 1769–70. *American Philosophical Society Transactions* 1 (1771): plate 7, fig. 1 (James Smithers, engraver).

nation, the Tulpehocken-Swatara connection seemed promising.

This trio of routes made up the Philadelphia plan for internal improvements, and they dominated the plans of those who sought to maintain control of Pennsylvania's lucrative hinterland. All these routes were in fact attempted and, with more or less delay, brought to completion, but not for decades. The visionaries of the pre-Revolutionary period had to be patient.

If the Philadelphians of 1768–72 had the wit and enthusiasm to identify proper routes (one Philadelphian thought the city likely to become "canal mad"),[17] they also recognized that they did not possess the skill to execute the projects. For

[17] Samuel Rhoads to Benjamin Franklin, 3 May 1771, in *The Papers of Benjamin Franklin,* ed. Leonard W. Labaree and William B. Willcox et al., 23 vols. in progress (New Haven and London: Yale University Press, 1959–), 18: 93.

such expertise Philadelphians naturally looked abroad. The polytechnic philosopher Benjamin Franklin was in England as the agent of the Pennsylvania Assembly during this time, and Philadelphians asked him for advice and reflection.

Thomas Gilpin had written to Franklin with high hopes after the survey for a Chesapeake-Delaware canal, but conceded that his activities were probably "more zealous than useful."[18] In 1771 Samuel Rhoads, a wealthy builder and politician who was a vice-president of the American Philosophical Society, sent Franklin a request for the latest "accounts and instructions respecting canals" and their construction.[19] A year later he thanked Franklin for sending "the reports of the great engineers, Smeaton, Brindly &c" which contained "a great deal of instruction to us inexperienced Americans." Rhoads had been on the survey with Rittenhouse, and was encouraged by the apparent practicality of a canal on the Tulpehocken-Swatara route. But, he asked Franklin, is it better to dig a canal, or just to dam up the rivers and creeks to provide for navigation?[20]

Franklin by this time had traveled on Dutch canals, had toured the Duke of Bridgewater's Canal, and had met and talked with the experienced engineer John Smeaton, so he could give an unequivocal reply.[21] He asserted that digging a canal was clearly the choice of experienced men, for

> rivers are ungovernable things, especially in hilly countries; canals are quiet and very manageable: therefore they are often carried on here by the sides of rivers, only on ground above the reach of floods, no other use being made of rivers than to supply occasionally the waste of water in the canals.

Moreover, Franklin pointed out, the business of canalling was best done by a competent civil engineer, and those promoters who had proceeded without one had paid dearly for their error. He urged that "if any work of that kind is set on foot in America, I think it would be saving money to engage by a

[18] Thomas Gilpin to Benjamin Franklin, 10 October 1769, ibid., 16: 216.

[19] Samuel Rhoads to Benjamin Franklin, 3 May 1771, ibid., 18: 94; s.v., "Samuel Rhoads," in *Appleton's Cyclopaedia of American Biography*, ed. James Grant Wilson and John Fiske, 6 vols. (New York: D. Appleton and Company, 1888–89).

[20] Samuel Rhoads to Benjamin Franklin, 30 May 1772, *Franklin Papers*, 19: 157–58.

[21] Ibid., 15: 115–16, 18: 114–15, 20: 476.

handsome salary an engineer from hence who has been accustomed to such business."[22] Rittenhouse had undoubtedly read Franklin's letter before he made his final report on the Tulpehocken-Swatara line, and he also recommended "procuring from *Europe* such assistance as the importance of the work may require."[23]

Franklin's advice was sound, but Rittenhouse's words came early in 1773, and the events leading to the Revolution had begun to divert attention from internal improvements. Securing an English engineer grew increasingly unlikely, and the canal and road projects lay dormant.

* * *

Almost immediately after the Peace of Paris took effect in 1783 the Pennsylvania Assembly called for another survey of the Tulpehocken-Swatara route, again asking Rittenhouse to lead a team of surveyors. While reaffirming the feasibility of the route, the report written in 1784 reveals by its terminology the surveyors' lack of familiarity with canal technology and does not indicate a significant advance over the knowledge of a decade earlier.[24] To move the Philadelphia plan for internal improvements from the stage of a viable conception to a viable technology now required the transfer of European civil engineering to the United States, as well as the creation of private or government financing sufficient to carry out the projects.

But for no apparent reason, Philadelphians of the immediate post-Revolutionary period hesitated, and the first post-Revolutionary canal projects of significant size were set in motion in the South. The Santee Canal in South Carolina, the James River Canal in Virginia, the Potomac River navigation of Virginia and Maryland, and the Susquehanna Canal in Maryland were all initiated in the mid-1780s.[25] Each soon

[22] Benjamin Franklin to Samuel Rhoads, 22 August 1772, ibid., 19: 278–79.

[23] Hindle, *David Rittenhouse,* 95.

[24] Ibid., 250–51; draft of a report to the Pennsylvania Assembly by David Rittenhouse, Thomas Hutchins, and Nathan Sellers, [May–June] 1784, Peale-Sellers Papers, American Philosophical Society (hereafter APS), Philadelphia, Penna.

[25] Taylor, *The Transportation Revolution,* 32; Christopher T. Baer, ed., *Canals and Railroads of the Mid-Atlantic States, 1800–1860* (Greenville, Del.: Regional Economic History Research Center, 1981), 52, 54–55.

felt the lack of a suitably qualified person to direct construction, and as Daniel Calhoun has pointed out, each limped along unsatisfactorily with a series of native managers and craftsmen, as well as immigrants who claimed some technical skill.[26]

George Washington was one of the most active canal promoters of the period, and he probably came to grips with the need for engineering skill with greater insight than most. Trained as a surveyor and having experience with talented French military engineers during the Revolution, Washington knew there was a difference between a craftsman's skill and professional engineering.[27]

In 1785, just after Maryland and Virginia had authorized formation of the Potomac Company, Washington wrote to Lafayette (who had returned to France) that "if a company should be established and the work is undertaken, a skillful Engineer, or rather a person of practical knowledge will be wanted to direct and superintend it."[28] Lafayette wrote back to tell Washington that he could obtain a proper French engineer from the government's corps of bridges and highways for 500 guineas a year.

Washington's reaction was that the cost was too high, but he made the same observation about engineering works that Franklin had earlier: "to begin well . . . is all in all: error in commencement will not only be productive of unnecessary expense, but, what is still worse, of discouragements." Washington suggested resolving the problem of the high salary by an imaginative plan: "It appears to me . . . that if the cost of

[26] Daniel Hovey Calhoun, *The American Civil Engineer: Origins and Conflict* (Cambridge, Mass.: The Technology Press of M.I.T. Press, 1960), 8–13.

[27] Paul K. Walker, ed., *Engineers of Independence: A Documentary History of the Army Engineers in the American Revolution, 1775–1783* (Washington, D.C.: Office of the Chief of Engineers, 1981), 10–22, 34–43. Washington made a trip to western Virginia and western Pennsylvania in September and October 1784, partly "to obtain information of the nearest and best communication between the Eastern and Western Waters; & to facilitate as much as in me lay the Inland Navigation of the Potomak; . . . ." The next year he was reported to have talked for two days about little else besides canals. Donald Jackson and Dorothy Twohig, eds., *The Diaries of George Washington,* 6 vols. (Charlottesville: University Press of Virginia, 1976–79), 4: 1–71 (quote on p. 4); Elkanah Watson, *Men and Times of the Revolution,* 2nd ed., reprinted (New York: Crown Point Press, 1968), 281.

[28] George Washington to Marquis de Lafayette, 15 February 1785, in *The Writings of George Washington,* ed. John C. Fitzpatrick, 39 vols. (Washington: Government Printing Office, 1931–44), 28: 73.

bringing from Europe a professional man of tried and acknowledged abilities, is too heavy for one work; it might be good policy for several Companies to unite in it . . . one man may plan for twenty to execute."[29] Washington did not succeed in obtaining an engineer, but his scheme foreshadowed an important aspect of the rebirth of the Philadelphia plan.

The new stirrings of interest in internal improvements in Philadelphia may be traced to a remarkable conjunction of individuals and institutions in Philadelphia in the early 1790s. Not the least important was the arrival of the new federal government in 1790, bringing not only a new optimism about the promotion of commerce, but numerous individuals with enthusiasm for and knowledge about internal improvements. Washington has been mentioned. His vice-president, John Adams, had been on Dutch canals, and Secretary of State Thomas Jefferson had spent several days examining the Canal du Midi in France. Congress had canal promoters in W. L. Smith from South Carolina and Philip Schuyler of New York, both of whom had seen European canals. The venerable Franklin was living in Philadelphia until his death in 1790.[30] All these men were members of the American Philosophical Society.

The immediate impulse for the revival of the Philadelphia plan came from a group of Philadelphians, four of whom were clearly central to the effort, perhaps because they had greater interests in technology than the others. Two were men already noted: David Rittenhouse, surveyor and instrument-maker, and William Smith, who, possibly because he had surveying talents and had been to England, was later asked to serve on numerous canal committees of oversight and investigation. By 1790 Smith was involved in Pennsylvania land speculation

[29] George Washington to William Moultrie, 25 May 1786, ibid., 28: 439–40.

[30] Lyman H. Butterfield, ed., *The Adams Papers.* Series I: *Diary and Autobiography of John Adams,* 4 vols. (Cambridge, Mass.: The Belknap Press of Harvard University Press, 1961), 2: 456, George C. Rogers, Jr., *Evolution of a Federalist: William Loughton Smith of Charleston, 1758–1812* (Columbia: University of South Carolina Press, 1962), 131–34; *Biographical Dictionary of the American Congress, 1774–1927* (Washington, D.C.: Government Printing Office, 1928), 1544; s.v., "Philip Schuyler," in *The Dictionary of American Biography* (hereafter *DAB*), eds. Allen Johnson and Dumas Malone, 20 vols. (New York: Charles Scribners' Sons, 1928–36); Julian P. Boyd, ed., *The Papers of Thomas Jefferson,* 19 vols. in progress (Princeton: Princeton University Press, 1950–), 11: 371–72, 446–54.

and Rittenhouse had explored much of the state as a member of boundary survey teams.[31]

Rittenhouse and Smith were intertwined by friendship and finance with Robert Morris and John Nicholson, the giants of real estate speculation in the era. Their investments included lands as far apart as Georgia, Kentucky, western New York, and the District of Columbia, but focused on central and western Pennsylvania.

Robert Morris was born in England in 1738 and had come to Maryland at age thirteen with his father, who became a tobacco agent. Morris eventually went to work for a Philadelphia merchant, and in 1754 became a partner in a commission business. He sometimes traveled abroad with his cargoes and at some point was in Europe. During the Revolution he became the most effective financial agent of the American government, and carried on a substantial and lucrative private business as well. He later used his knowledge of government to become a major purchaser of public lands. He became an intimate of Washington's and rented his house to Washington while he was president. Morris then lived next door and saw Washington frequently both as a friend and as a senator in the new Congress. Morris had schemes for a massive industrial site at the falls of the Delaware River, where he had a considerable ironworks under construction in the early 1790s.[32]

John Nicholson was born in Wales in 1757 and emigrated with his parents to Pennsylvania at an early age. In the Revolution Nicholson became a prominent official of the state government, and, like Morris, emerged with a genius for real estate manipulation. Nicholson developed an industrial complex at the falls of the Schuylkill and was a close friend of

[31] s.v., "David Rittenhouse," and "William Smith," *DAB*; Hindle, *David Rittenhouse,* 218, 255–56, 262–70, 320–21; William Smith cashbook and "Miscellaneous bills, receipts, accounts file," in Miscellaneous A–M box, William Smith Section (hereafter WSS), Jasper Yeates Brinton Collection, Historical Society of Pennsylvania (hereafter HSP), Philadelphia, Penna. I make my comments on Smith's technical services as a result of examining records of the Schuylkill and Susquehanna Navigation Company, the Delaware and Schuylkill Navigation Company, and the Conewago Canal Company.

[32] Ellis Paxon Oberholtzer, *Robert Morris: Patriot and Financier* (New York: The Macmillan Company, 1903), 1–9, 276–77, 295–96; George Washington to Robert Morris, 27 March 1790, item 1636, Sol Feinstone Collection of the American Revolution, APS; s.v., "Robert Morris," *DAB.* There is no recent authoritative biography of Morris.

Rittenhouse and Smith.[33] Again, the esteem in which these four men were held by the technically inclined intellectuals of Philadelphia was indicated by their election to the American Philosophical Society.[34]

Morris, Nicholson, Smith, and Rittenhouse appear to have been prime movers in the creation of a lobbying group titled the Society for the Improvement of Roads and Inland Navigation, which began meeting in January 1791.[35] Numerous merchants and respectable mechanics soon joined. The Society resolved to meet weekly while the state assembly was in session in Philadelphia, believing that the legislators had in mind "an extended plan for improving roads and amending inland navigation." Pennsylvania had just adopted a new constitution with a more statist tone than the previous document, and the time apparently seemed right for a Hamiltonian approach to the promotion of commerce.[36]

Robert Morris was elected the society's president, and the group's first act was to issue under his name a memorial to the assembly written by a committee including Smith and Nicholson. The memorial argued the same theme heard twenty years earlier: if Philadelphia did not take control of the trade of the interior "the loss may never be retrieved," and that transportation improvements were needed to make the state's lands attractive to settlers. The memorial offered a detailed chart of distances along major watercourses to show that Philadelphia (and Pittsburgh) were the closest ports to various points in the interior of Pennsylvania. Moreover, a canal over the Tulpehocken-Swatara route was emphasized as the critical link in any system of internal navigation, and an estimate of its cost was based upon a special survey carried out two months before. The memorial gave special attention to the increasing grain trade on the Susquehanna as a source

[33] Robert D. Arbuckle, *Pennsylvania Speculator and Patriot: The Enterpreneurial John Nicholson, 1757–1800* (University Park and London: The Pennsylvania State University Press, 1975), 5, 40–41, 145; Hindle, *David Rittenhouse,* 239.

[34] Membership card file, APS.

[35] Preamble, Journal of the Society for the Improvement of Roads and Internal Navigation (hereafter Jol. SIRIN), photostats, collection 1085, HSP. Various authors have stated that the society began meeting in 1789, an error which probably originated in Smith, *Historical Account,* p. xvi.

[36] Jol. SIRIN, preamble, 31 January 1791, 3 February 1791, 14 February 1791, 21 February 1791; Stevens, *Pennsylvania,* 118, 120.

of income for the proposed canal.[37] The society followed this memorial with a request that the state spend $300,000 for internal improvements, but the state's actual appropriation in April 1791 included amounts for road and river improvements rather than canals.[38]

The society then petitioned the assembly to form a corporation to make the canal, and on 27 September 1791 they complied with the society's wishes, creating the Schuylkill and Susquehanna Navigation Company.[39] Stock subscriptions to the new company lagged somewhat, and in November Morris even convened a special meeting of the Society for the Improvement of Roads and Internal Navigation "to take into consideration the measures necessary to give a spring to the subscriptions."[40] A few days later the stock commissioners reported the requisite number of subscriptions, and in January 1792 the stockholders elected Robert Morris the company's president, and placed William Smith and John Nicholson among the twelve managers.[41]

Morris had already taken the initiative of writing to Patrick Colquhoun, a correspondent in Britain, to ask him to recruit an engineer for the infant company.[42] It seems likely that he did not take this step on his own, but probably acted on the advice of his friends Washington, Smith, Nicholson, and Rittenhouse.

Moreover, Morris may have recognized the acuteness of the need for an experienced technician upon meeting the self-declared engineer James Brindley, the best America had to offer at the time. Brindley was a British immigrant who had worked on the Susquehanna Canal in Maryland in the 1780s and who as late as 1799 was engaged on various projects in

[37] Jol. SIRIN, 31 January 1791, 7 February 1791; Smith, *Historical Account*, 1–22; Timothy Matlack, Samuel Maclay, and John Adlum, "Digging a Canal from Kucher's Mill-dam to Myers Mill-dam," copy, 14 December 1790, Union Canal Papers, collection 1328, HSP.

[38] Jol. SIRIN, 14 March 1791; Smith, *Historical Account*, 73–75.

[39] Smith, *Historical Account*, 23–32.

[40] Jol. SIRIN, 30 November 1791.

[41] Minutes of the Proceedings of the President, Managers, and Company of the Schuylkill and Susquehanna Navigation (hereafter S and S mins.), 9 January 1792, Union Canal papers.

[42] Colquhoun's first response to Morris was in a letter dated 4 January 1792. Allowing for transatlantic travel, Morris had probably written to him two months earlier. S and S mins., 2 April 1792.

the United States. The Society for the Improvement of Roads and Internal Navigation had called on him in February 1791 to examine the Tulpehocken-Swatara route. Brindley came with his leveling instrument a few days later, and with a committee of the society carried out a survey. Although Brindley did what was asked of him, Morris probably had reason to agree with Washington's earlier lukewarm assessment that Brindley had only "more *practical* knowledge of cuts and locks for the improvement of inland Navigation, than any man among us."[43]

In any case, Morris and the company decided not to rely on Brindley,[44] but to try to acquire a full-fledged British engineer. The first hopeful report from Colquhoun came in April 1792 when he notified Morris that he expected to sign the engineer Thomas Dadford, Jr., to a contract to come to America. Morris wrote back to urge Colquhoun to act with expedition, but in September came word that Dadford had decided not to come.[45]

It was now a year since the company had been chartered, and although still another but more detailed survey of the projected canal route had been done by a team of surveyors headed by William Smith, the company had yet to turn the first spadeful of earth. The reason for this inactivity became obvious at a meeting of the company on 25 September 1792. According to the minutes:

> The Board went into a consideration of the various modes proposed for . . . a canal across the middle ground. . . . Also the propriety of beginning the navigation on the Tulpehocken and

[43] Jol. SIRIN, 21 February 1791, 28 February 1791, 7 March 1791; S and S mins., 16 January 1792, 3 March 1792; George Washington to William Moultrie, 25 May 1786, *The Writings of George Washington,* 28: 439; Merrit Roe Smith, *Harpers Ferry Armory and the New Technology: The Challenge of Change* (Ithaca: Cornell University Press, 1977), 39, 41–42.

[44] The company did hire him for part of the summer of 1792 to conduct more surveys, but made clear that is was for "the season" only. Brindley soon was hired as resident engineer for the Conewago Canal, for which see below: S and S mins., 2 April 1792, 9 April 1792, 21 May 1792, 11 June 1792.

[45] Ibid., 2 April 1792, 30 April 1792, 10 September 1792, 18 September 1792; Charles Hadfield, *British Canals: An Illustrated History,* 4th ed. (New York: Augustus M. Kelley, 1969), 42; s.v., "Patrick Colquhoun," in *Dictionary of National Biography,* ed. Leslie Stephen and Sidney Lee, 22 vols. (Oxford: Oxford University Press, 1885–1901); A. W. Skempton and Esther C. Wright, "Early Members of the Smeatonian Society of Engineers," *Transactions of the Newcomen Society* 44 (1971–72): 36.

> Quitapahilla and Swatara, or each of them—and whether the bed of the river shall be followed or a canal dug along the side of the bank—but no question was put.[46]

They needed an engineer to bring an end to such confusion over technical matters.

Fortunately, at the end of October Morris received a letter from Colquhoun which announced that a firm contract had been concluded with another British engineer, William Weston, and that Weston could be expected in Philadelphia by Christmas. With a sense of relief the managers voted their thanks to Colquhoun for "his unremitted attention in procuring an able and experienced engineer."[47] We do not know what Colquhoun told them about Weston which gave them such confidence in his choice, but what is known of Weston's background indicates that Colquhoun could have assured them that he fit their needs.

Weston, a resident of Oxford, was twenty-nine at the time he accepted his overseas commission. He had obtained engineering experience on the construction of the Oxford Canal, then in 1787 was selected to direct the erection of a major bridge over the Trent River at Gainsborough. William Jessop, the eminent engineer who was currently the consultant on the Trent Navigation, helped plan the bridge. It was undoubtedly with his consent that Weston was chosen as the constructing engineer, and in 1791 Weston successfully completed the three-arched stone bridge of 328 feet. Jessop must have recognized the quality of the work, for an article published in the next edition of the *Encyclopedia Britannica* reported that when Jessop was asked "to select a properly qualified engineer" for North America, he recommended Weston.[48]

[46] S and S mins., 25 September 1792. The Quitapahilla is a tributary of the Swatara Creek.

[47] Ibid., 31 October 1792.

[48] Charles Hadfield and A. W. Skempton, *William Jessop, Engineer* (North Pomfret, Vt.: David & Charles, 1979), 25–27, 30; *The Edinburgh Encyclopedia,* 1st American ed. of the *Encyclopedia Britannica,* 21 vols. (Philadelphia: Joseph and Edward Parker, 1832), s.v., "William Jessop"; Ian Beckwith, "Gainsborough: The Industrial Archeology of a Lincolnshire Town," *Industrial Archeology* 8 (1971): 268; *The History of the County of Lincoln,* 2 vols. (London and Lincoln: John Saunders, 1834), 2: 17; notebook of William Weston, 624/629 (093.32), Library of the Institution of Civil

In accepting the challenge, Weston also found a companion with whom to share it. On 23 October 1792 he married Charlotte Whitehouse of Gainsborough,[49] and a month later they embarked on a vessel for a spectacular honeymoon—eight years in the United States.[50]

Weston presented himself and the contract executed by Colquhoun at a board meeting of the Schuylkill and Susquehanna Navigation Company in mid-January 1793. The contract required him to be at the disposal of Morris or the company for seven months out of every year for five years, with a salary of £500 per year. The remaining five months of each year Weston could use as he pleased.[51] There is little doubt that Morris had set these terms, especially the clause allowing Weston his own time, for Morris had earlier written to the governor of Pennsylvania proudly referring to the imminent arrival of an engineer "from whose abilities we hope advantages will be derived, not only to [the Schuylkill and Susquehanna Navigation] but generally to the inland navigation of America."[52]

Engineers, London. Weston's notebook has numerous references to the Oxford Canal and the Gainsborough Bridge.

Regarding Jessop's involvement in contacting Weston it is interesting that Franklin had met Jessop in 1773 and may have met with him later, perhaps informing him of America's need for a civil engineer: *Franklin Papers,* 20: 476; *The Edinburgh Encyclopedia,* s.v., "William Jessop."

The major sketch of Weston's life and work remains Richard Shelton Kirby's "William Weston and His Contribution to Early American Engineering," *Transactions of the Newcomen Society* 16 (1935–36): 111–27. Elting E. Morison also has written about Weston and has effectively developed the significance of Weston's skills for American technology. Unfortunately he does not sufficiently identify his sources and makes several dubious assertions about Weston's career, such as stating that he had his training under the great British engineer James Brindley. Since Weston was 29 years old in 1792, he would have been only nine at Brindley's death in 1772. Elting E. Morison, *From Know-How to Nowhere: The Development of American Technology* (New York and Scarborough, Ont.: New American Library, 1974), 22–27, 31–33, and next note, below.

[49] Nigel Colley, private communication to author, 26 September 1983. Mr. Colley cites the transcript of the Gainsborough parish registers for the marriage date, and notes that "the bond and allegation for the marriage license (MB 1792/723-4) describe Weston as an engineer, aged 29." Later Weston spoke of giving his wife "a solemn promise . . . previous to obtaining her consent to accompany me to America," apparently a promise to return to England if needed by her family. William Weston to Richard Peters, 4 May 1803, Society Miscellaneous Collection, HSP.

[50] S and S mins., 22 January 1793.

[51] Ibid., 14 January 1793.

[52] Robert Morris to Thomas Mifflin, 29 November 1792, in S and S mins., 29 November 1792.

Morris must have had in mind primarily three other internal improvement companies of which he was also a major promoter, and whose boards of managers were interlocked with the Schuylkill and Susquehanna Company: the Delaware and Schuylkill Navigation, the Philadelphia and Lancaster Turnpike, and the Conewago Canal.[53] Each of these companies also was born out of the efforts of the Society for the Improvement of Roads and Internal Navigation.

The Delaware and Schuylkill Company was chartered in the spring of 1792 to construct a canal from the Schuylkill to the Delaware River. The company, of which Morris was also president, laid out a route which began at Norristown, followed the Schuylkill to near Fairmount, then turned east to meet the Delaware. The company's charter permitted it to sell some of its water to the citizens of Philadelphia for culinary and sanitary use.[54]

The state assembly also created the Philadelphia and Lancaster Turnpike Company in the spring of 1792, and its shares were enthusiastically subscribed. The road was surveyed by December 1792, and essentially followed the existing road between the cities. Following two years of construction it became the first and one of the most successful of the new republic's turnpike companies.[55]

The last company chartered was the Conewago Company, formed to build a one-mile canal around a set of dangerous falls on the Susquehanna River between the mouth of Swatara Creek and Wright's Ferry. In 1792 Robert Morris proposed that the Schuylkill and Susquehanna Company undertake the work, but the managers rejected his proposal. Instead, Morris, Smith, and Nicholson worked out a plan whereby the governor of Pennsylvania would allocate some of the state's internal improvement monies to them as contractors for the

[53] Smith, *Historical Account,* 1, 44; Hindle, *David Rittenhouse,* 322–23. I have not found a list of the Philadelphia–Lancaster Turnpike directors, but the first president was William Bingham, who was on the other three boards: William Bingham to Thomas Mifflin, 1 December 1792, William Bingham Letterbook, collection 53, HSP.

[54] Smith, *Historical Account,* 33–44.

[55] Joseph A. Durrenberger, *Turnpikes: A Study of the Toll Road Movement in the Middle Atlantic States and Maryland* (Valdosta, Ga.: Southern Stationery and Printing, 1931), 51–52; William Bingham to Thomas Mifflin, 1 December 1792, William Bingham Letterbook, HSP.

project. In July, along with several other Schuylkill and Susquehanna managers and David Rittenhouse, they agreed "to open and improve the navigation," for the sum of £5,250.[56] This group hired James Brindley to supervise the work, but little appears to have been done until after the spring of 1793 when the contractors resolved themselves into the Conewago Canal Company under a state charter.[57]

These three companies attempted to contract for the remainder of Weston's time—the five months reserved for Weston's own use. In some complex negotiations not fully revealed by the records Weston agreed to bind himself for the remainder of his time to the Schuylkill and Susquehanna Company, which then granted him absences to work for the other companies. Almost immediately after this agreement was reached Weston began to share his knowledge with the Delaware and Schuylkill, Conewago, and Turnpike companies.[58]

But Weston's central responsibility was to the Schuylkill and Susquehanna Canal, which the promoters of the Philadelphia plan called the "golden link" to the west. Upon his arrival Weston found that the company's managers had attempted a cautious start on the project. Convinced that surveys by Rittenhouse and Smith were accurate, in October 1792 they chose a route between the headwaters of the Tulpehocken and the Quitapahilla, a tributary of the Swatara, and set men to excavating a canal. By the end of November nearly half a mile was dug near Lebanon and about a hundred men were at work.[59]

Weston made his first trip to the canal in early February, two weeks after presenting himself to the company for the first time. Accompanied by William Smith, Weston examined the route of the canal and reported back to the managers.[60]

[56] S and S mins., 26 June 1792, 28 June 1792; *Pennsylvania Archives,* 9th ser., 1: 421; Thomas Mifflin to Robert Morris, William Smith, and John Nicholson, 29 June 1792, Conewago Canal folder, Miscellaneous A–M box, WSS.

[57] Smith, *Historical Account,* 44–47, 67–68. In February 1794 Brindley reported work on the Conewago Canal was going slowly. James Brindley to John Nicholson, 25 February 1794, General Correspondence, John Nicholson Papers (hereafter JNP), Pennsylvania Historical and Museum Commission, Harrisburg.

[58] S and S mins., 16 January 1793, 22 January 1793, 26 February 1793, 24 March 1793, 26 March 1793.

[59] Ibid., 8 October 1792, 29 November 1792.

[60] Ibid., 19 February 1793; Smith, *Historical Account,* 68.

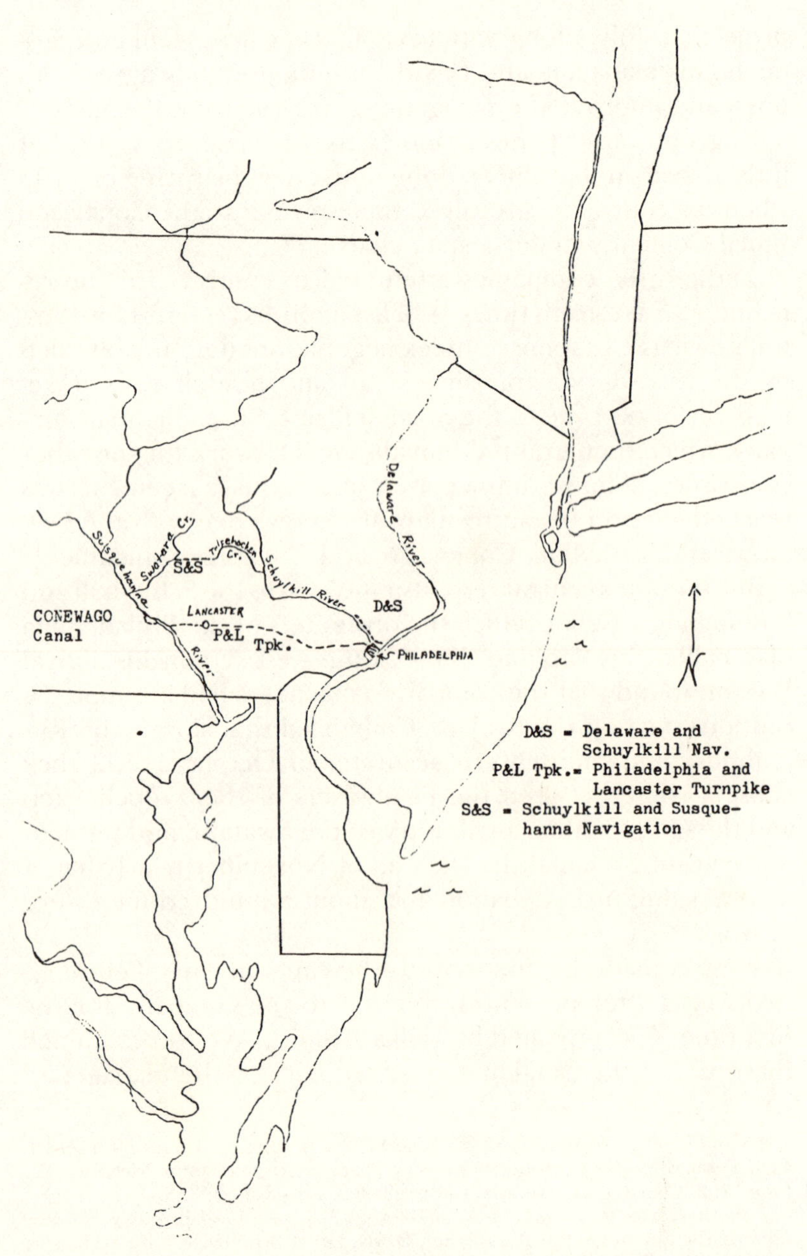

Fig. 2. Map of the four associated internal improvement companies in Pennsylvania, 1792.

Diplomatically he began his report by affirming that they had chosen the correct route for the summit level of the canal, but thereafter his comments clearly reflected his claims to superior technical knowledge. He made recommendations regarding the width and depth of the canal, the towpath, and the locks. He proposed that:

> all things touching the works, the Superintendants or Overseers, the workmen, and all persons contracting with the Company for the execution of any part or parts of the same, be under the direction of Mr. Weston, the Engineer, according to such instructions as he may receive from the President and Managers.[61]

He asked for committees to assist him in surveying and in purchasing lands. He wanted the authority to organize the workmen into companies (a practice on British canals), to set regulations, and to hire and fire workmen and overseers. He asked for a contract to be made for spades, probably on a pattern he had brought from England, and he asked to be empowered to make bricks and burn lime.[62] With this display of knowledge and authority Weston took charge of operations.

He did not return to the scene of excavation for some time, devoting a month and a half to the turnpike and the Delaware and Schuylkill Canal before removing to Lebanon with his wife. Even then he was called to the Conewago Canal, and subsequently the terrible yellow fever epidemic in Philadelphia during the summer and fall must have been disruptive. The year of 1793 seems, as a result, to have been largely a time when Weston carried out the organizational framework he had formulated in February, although he did supervise the

[61] Weston's first report does not survive in its entirety. An extract is printed in Smith, *Historical Account,* 68, and the substance of it is in S and S mins., 26 February 1793.

[62] S and S mins., 26 February 1793. Morison, *From Know-How to Nowhere,* 24, mentions that Weston "lent out . . . his spade," a bit of information probably gleaned from the Baldwin-Weston letters, which I have not seen. On British canal labor practice, see Anthony Burton, *The Canal Builders,* 2nd ed. (Newton Abbot: David and Charles, 1981), 156, 159. For the special shape of the spade used for digging canals, see Darwin H. Stapleton, ed., *The Engineering Drawings of Benjamin Henry Latrobe* (New Haven and London: Yale University Press, 1980), 16, fig. 8.

digging of about three more miles of canal, and the building of brick and lime kilns, and a sawmill.[63]

More importantly, however, in the fall Weston fulfilled the managers' instruction that he examine the Tulpehocken down to its junction with the Schuylkill at Reading, with a view to improving it for navigation. His report carried an emphasis which echoed the comment of Franklin twenty years before:

> The contest between river navigation and canals is an old one. . . . The unerring test of experience has at length convinced the warmest advocates for river navigations how inefficacious they are. . . . As far as my opinion will influence the board . . . I recommend . . . the adoption of a canal navigation. . . .[64]

Weston recognized the impact of his report on the plans of the company, which had been formed with the idea of digging a summit canal of about ten miles and perhaps clearing some difficult places on connecting rivers. He considered digging a canal along thirty-four miles of the Tulpehocken and building the forty-five locks required, and estimated that it would take four years and cost about £220,000. Adding in the probable cost of improving the Swatara (he promised a survey the next summer), and the money already expended by the company, Weston came to an estimate of £458,000 for completing the work. This, he noted, was three times the capitalization which the state assembly had permitted the company. Optimistically, he argued that completing just the segment from Lebanon to Reading would result in considerable toll-generating traffic, which would reduce the financial burden.[65]

What Weston did not note, but which the managers were painfully aware of, was that many of the subscribers to the Schuylkill and Susquehanna Company's stock had become delinquent in their payments, and the company was already headed for financial trouble. Upon receiving Weston's report in February 1794, the managers tried a number of avenues to obtain more capital.

[63] S and S mins., 26 March 1793, 24 May 1793, 5 September 1793, 3 January 1794, 25 March 1794; J. H. Powell, *Bring Out Your Dead: The Great Plague of Yellow Fever in Philadelphia in 1793* (Philadelphia: University of Pennsylvania Press, 1949); Smith, *Historical Account,* 57.

[64] Smith, *Historical Account,* 51–52: S and S mins., 3 February 1794.

[65] Smith, ibid., 49, 57.

Robert Morris first addressed an appeal to the state assembly, reminding the legislators that "the canal which is to connect the *Schuylkill* and *Susquehanna* is the chief link" in the proposed system of state internal improvements. He cited Weston's report, and pointed out that the company now contemplated expending three times its original capitalization, a sum which might be raised by increasing the number of shares issued or by borrowing from the state. Given the current delinquencies, it is understandable that Morris, speaking for the managers, requested a loan. The legislators were unresponsive.[66]

By this time it was April, when British canal engineers usually expected to begin the year's campaign. The managers reviewed the company's finances, consulted with Weston, and considered that they had debts of $88,000, but only $110,000 on hand. Although there was sentiment to stop the work, the managers decided that course was "inconvenient" and "mischievious." They went to the extreme of mortgaging unpurchased shares, and tried to borrow from sympathetic stockholders.[67]

The previous season having been devoted in part to getting ready for a major effort, Weston was pleased that he could finally begin to carry out the work according to his plans. Despite difficulty in obtaining regular cash remissions from the company, by December Weston reported that he had doubled the excavated length of the canal. He had made nearly three million bricks and seventy-five thousand bushels of lime, and he had constructed five locks and two bridges. Materials were on hand to build several more.[68]

But the end was near. By the spring of 1795 the company had no more funds, and resorted to obtaining permission from the assembly to operate a lottery in order to raise money. In the meantime Weston was authorized to sell off materials and teams of animals, and to hire watchmen to protect the canal.[69] John Nicholson wrote to Weston hopefully, claiming that he

[66] S and S mins., 21 February 1794, 29 March 1794.
[67] Ibid., 2 April 1794, 3 May 1794, 7 May 1794.
[68] William Weston to John Nicholson, 19 June 1794, 20 September 1794 (copy), and Isaac Roberdeau to John Nicholson, 19 July 1794, 21 August 1794, General Correspondence, JNP; Smith, *Historical Account,* 58–61.
[69] S and S mins., 9 April 1975; *Pennsylvania Archives,* 9th ser., 2: 979–80.

could still expect "a call in the present season . . . to resume the works at the Tulpehoccon," but the call never came.[70]

The company released Weston to go to other projects. He had, in fact, begun consulting on internal improvements outside Pennsylvania the previous summer when, with the Schuylkill and Susquehanna managers' permission, he had spent about two weeks in Massachusetts examining the proposed line of the Middlesex Canal and advising on matters of construction.[71] On his return he found that the managers of the Potomac Company wanted him to visit their works, but only after Washington personally asked for Weston's services in December was Weston released for a flying visit early in 1795.[72]

Thereafter Weston was largely employed by the New York Western Inland Lock Navigation, a company which for two years had been attempting to construct a canal to connect the Mohawk River with a tributary of Lake Ontario. Weston joined the company in the late spring of 1795, when the state of New York strengthened the company's finances by purchasing 200 shares of stock. For the next three years Weston directed the construction of two sets of locks on the Mohawk, and completed the excavation of a summit level canal near Rome, New York. Weston also surveyed the Mohawk River in order to collect data for a future canal along it.[73]

Weston consulted on the James River Canal in 1796, continued occasional work for the Delaware and Schuylkill Canal until 1798, drew up a report for a water supply system for New York City in 1799, and in 1800–01 assisted in planning for the Schuylkill Permanent Bridge in Philadelphia, which was by far the largest bridge built in the United States up to

[70] John Nicholson to William Weston, 15 April 1795, John Nicholson Letterbooks, collection 452, HSP.

[71] S and S mins., 29 May 1794; Christopher Roberts, *The Middlesex Canal, 1793–1860* (Cambridge, Mass.: Harvard University Press, 1938), 48–59; Morison, *From Know-How to Nowhere,* 22–25.

[72] S and S mins., 4 August 1794, 30 December 1794; George Washington to Tobias Lear, 21 December 1794 and 5 March 1795 in *The Writings of George Washington,* 34: 66, 132.

[73] S and S mins., 13 April 1795, 21 January 1796, 25 April 1796, 16 January 1797; Philip Schuyler, *Report of the Directors of the Western Inland Lock-Navigation Company, to the Legislature, 16th February, 1798* (Albany, N.Y.: Charles R. and George Webster, 1798), 4–16; Ronald E. Shaw, *Erie Water West: A History of the Erie Canal, 1792–1854* (Lexington: University of Kentucky Press, 1966), 13–20, 64.

that time.[74] But Weston remained mostly in upper New York after 1798, apparently content to live in what he termed "the paradise of this western country." Then late in 1801 Weston and his wife decided to return to England, and they left the United States permanently.[75]

Weston's departure did not deprive the Philadelphia promoters of technical expertise, for a second English engineer, Benjamin Henry Latrobe, was already in the United States. Latrobe had been born in England, and was raised and educated in the Moravian faith. About 1781 or 1782, when he was seventeen, he took an interest in architecture and engineering and began to acquire professional training in Germany and Britain. Apparently taken on as an assistant by the eminent engineers John Smeaton and William Jessop, he obtained experience on one canal and planned another. On arriving in the United States in 1796 he worked at a variety of commissions in Virginia before visiting Philadelphia in the spring of 1798 and being offered the opportunity to design and build the Bank of Pennsylvania there.

Samuel M. Fox, the bank's president, had served on a committee appointed by the Philadelphia city councils to consider the means of providing the growing American metropolis with a regular and clean water supply. When Latrobe moved to Philadelphia in December 1798 the committee chairman asked him to examine and report on a means of watering the city. Latrobe responded rapidly, and wrote a pamphlet which asserted that, by adopting the technology of the London waterworks with which he was familiar, the city could have its water supply by the summer.[76]

The impetus for building a waterworks was the yellow fever epidemic of 1793 which had recurred, though with less fe-

[74] Kirby, "William Weston," 121–24; Benjamin Henry Latrobe to James Wood, 14 February 1798, *Microfiche Ed.*; *Address of the Committee of the Delaware and Schuylkill Canal Company* . . . (Philadelphia: John Ormrod, [1797]), 7n–8n; extract of William Weston's report of 23 February 1796, in "The Agent for Real Estate &c. Reports," n.d., box 1, Union Canal Company of Pennsylvania Papers, RG 174, Pennsylvania Historical and Museum Commission, Harrisburg; Charlotte Weston to John Nicholson, 26 April 1796, General Correspondence, JNP.

[75] William Weston to William Meredith, 10 July 1800, and 10 October 1801, Meredith Papers, collection 1509, HSP.

[76] Stapleton, ed., *Engineering Drawings*, 3–8, 28; *Report of the Joint Committee of the Select and Common Councils, on the Subject of Bringing Water to the City* (Philadelphia: Zachariah Poulson, 1798).

rocity, every succeeding summer. Although there were differences of opinion among medical experts as to its cause, all seemed to agree that the incidence of disease could be reduced by drinking pure water and by regularly washing debris off the streets.[77]

The committee appointed by the Philadelphia city councils to look into this matter heard from a variety of persons regarding means of bringing water to the city, including the Delaware and Schuylkill Company, but Latrobe was the first to speak with authority. The committee ordered his report printed and immediately employed him as consulting engineer; just over two months later his modified plan to build a steam-powered waterworks was approved by city councils and Latrobe took charge of the greatest American civil engineering project of his day. Though he did not meet his goal of completing the waterworks in 1799, when put into operation in February 1801 the Philadelphia Waterworks was a success. Latrobe's high level of knowledge and skill was apparent to all.[78]

Latrobe had come on the scene propitiously, for the vision of the Philadelphia plan for internal improvements was as yet unfulfilled. Two of the three routes to the west had been attempted, one successfully, one unsuccessfully, but one remained: the improvement of the lower Susquehanna combined with a Chesapeake-Delaware canal. Agitation for this route never ceased, but had been unfruitful in part because it required the cooperation of three states. Finally in 1799 Pennsylvania offered to clear its portion of the Susquehanna below the Conewago Falls if Maryland and Delaware would incorporate the Chesapeake and Delaware Canal Company.

After two years those conditions were met and Pennsylvania appropriated $10,000 for the Susquehanna improvement. The governor signed a contract for the work with Frederick Antes, an experienced surveyor and contractor who was also La-

[77] Stapleton, ed., *Engineering Drawings,* 28–29; *Pennsylvania Archives,* 9th ser., 2: 1297, 1301, 1316.

[78] Stapleton, ed., *Engineering Drawings,* 29–35; memorandum regarding Charles Taylor, n.d., box A, Thomas Pym Cope Collection, Haverford College, Haverford, Penna.; *Report of the Joint Committee,* p. 3; Minutes of the Select Council, 23 January 1799, Archives of the City and County of Philadelphia, City Hall, Philadelphia.

Fig. 3. Portrait of Benjamin Henry Latrobe by Rembrandt Peale, ca. 1816. Courtesy of Mrs. Gamble Latrobe.

trobe's uncle, and Antes appointed Latrobe his engineer and surveyor.[79]

Latrobe carried out his tasks in the late summer and fall of 1801, directing the clearing of a channel through all the dif-

[79] Stapleton, ed., *Engineering Drawings,* 12, 75–76.

ficult rapids by blowing up and clearing rocks and debris. At the same time he made a survey of the Pennsylvania segment of the lower Susquehanna, and during the winter drew a large watercolor map based on his data.

By the spring of 1802, in conjunction with a similar effort below the Maryland line, the forty miles of the Susquehanna below Columbia had been made much safer for those who wanted to take their rafts, arks, and keelboats to the Chesapeake during spring high water. Although this navigation was limited, and Latrobe estimated that it would take an additional $100,000 to make the river a convenient artery of commerce, the backwoods Pennsylvanians finally had a transportation route to the sea.[80]

Meanwhile the legislatures of Pennsylvania, Maryland, and Delaware struggled to coordinate the chartering of a canal company. Pennsylvania initiated the process in 1799, and Maryland soon followed, but not until February 1802 did Delaware act to allow formation of the company. Over a year was allotted for soliciting stock subscriptions, so the stockholders did not meet until May 1803 to elect a president and directors. Pennsylvanians had subscribed to the majority of the shares and ultimately provided nearly three-quarters of the capital. A leader of the company from the beginning was Joshua Gilpin, a Philadelphia merchant who had studied canals in England and who was the son of Thomas Gilpin, the promoter of a Chesapeake-Delaware canal over thirty years earlier. The other Pennsylvania representatives were also Philadelphia merchants.[81]

Some canal promoters inquired after William Weston even before the company was formed, and he professed a willingness to return to the United States to prosecute the work.[82] Latrobe lived in Philadelphia, however, and must have been

[80] Ibid., 76–79, 82–84, 89–109; Thomas C. Cochran, ed., *The New American State Papers. Transportation,* 7 vols. (Wilmington: Scholarly Resources, 1972), 1: 26; Livingood, *The Philadelphia-Baltimore Trade Rivalry,* 37–38.

[81] Ralph D. Gray, "The Early History of the Chesapeake and Delaware Canal: Part I, Early Plans and Frustrations," *Delaware History* 8 (March 1959): 231–43; Stapleton, ed., *Engineering Drawings,* 12; Cochran, ed., *The New American State Papers. Transportation,* 1: 68.

[82] William Weston to Richard Peters, 4 May 1803, Society Miscellaneous Collection, HSP.

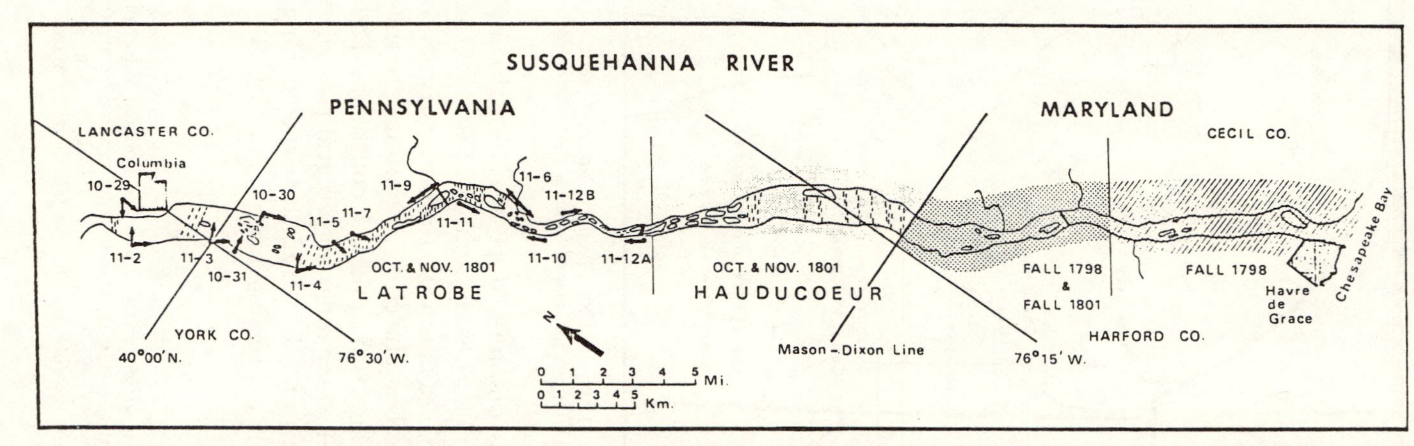

Fig. 4. Latrobe carried out his survey of the Susquehanna River from Columbia to the Maryland line in October–November 1801. Christian Hauducoeur, engineer of the Susquehanna Canal in Maryland, conducted a survey on the lower river simultaneously, supplementing his mapping of three years earlier. Stephen F. Lintner.

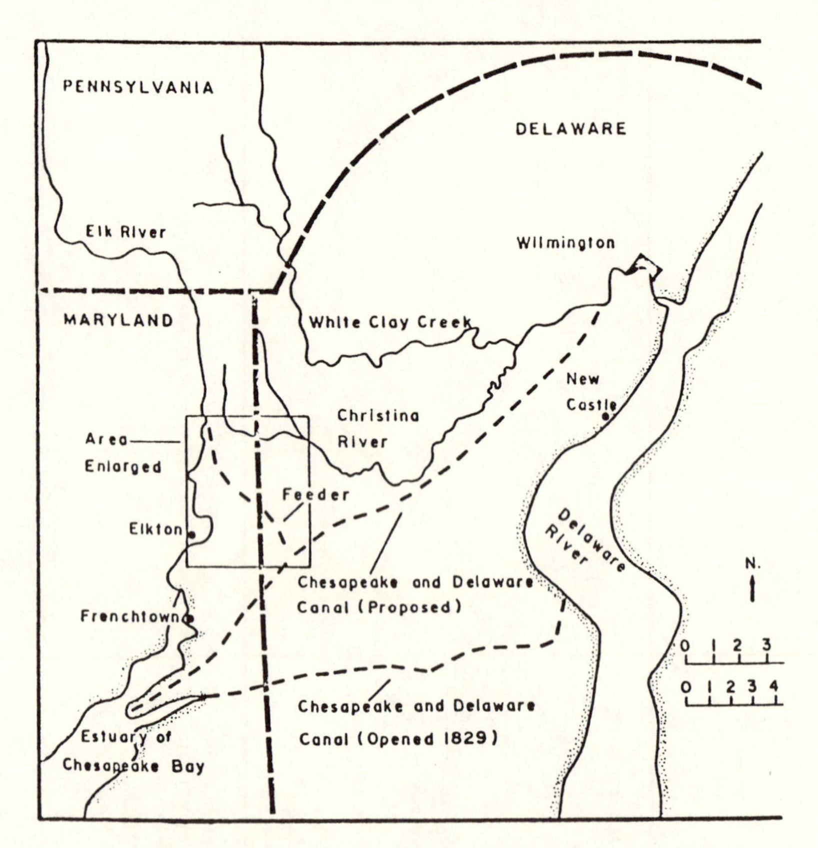

Fig. 5. Chesapeake and Delaware Canal feeder. Construction of the feeder lasted from 1804 to 1805, but for lack of funds it was not completed.

known to all the promoters. Moreover, they already had some idea of his waterway interests: in August 1799 he had examined the Delmarva peninsula to consider canal routes, that December he had been hired as a consultant to the Delaware and Schuylkill Company, and, of course, in 1801 he had worked on the Susquehanna, an integral part of the Chesapeake-Delaware route. In 1802 he wrote a long letter to Thomas Jefferson expressing his opinions on the route for the canal.[83]

[83] Edward C. Carter II, ed., *The Journals of Benjamin Henry Latrobe, 1799–1820: From Philadelphia to New Orleans* (New Haven and London: Yale University Press, 1980), 8; extract from the records of the Delaware and Schuylkill Navigation Company, 29 November 1799, box 1, Union Canal Company of Pennsylvania Papers; Benjamin Henry Latrobe to Thomas Jefferson, 27 March 1802–1 June 1802–24 October 1802, *Microfiche Ed.*

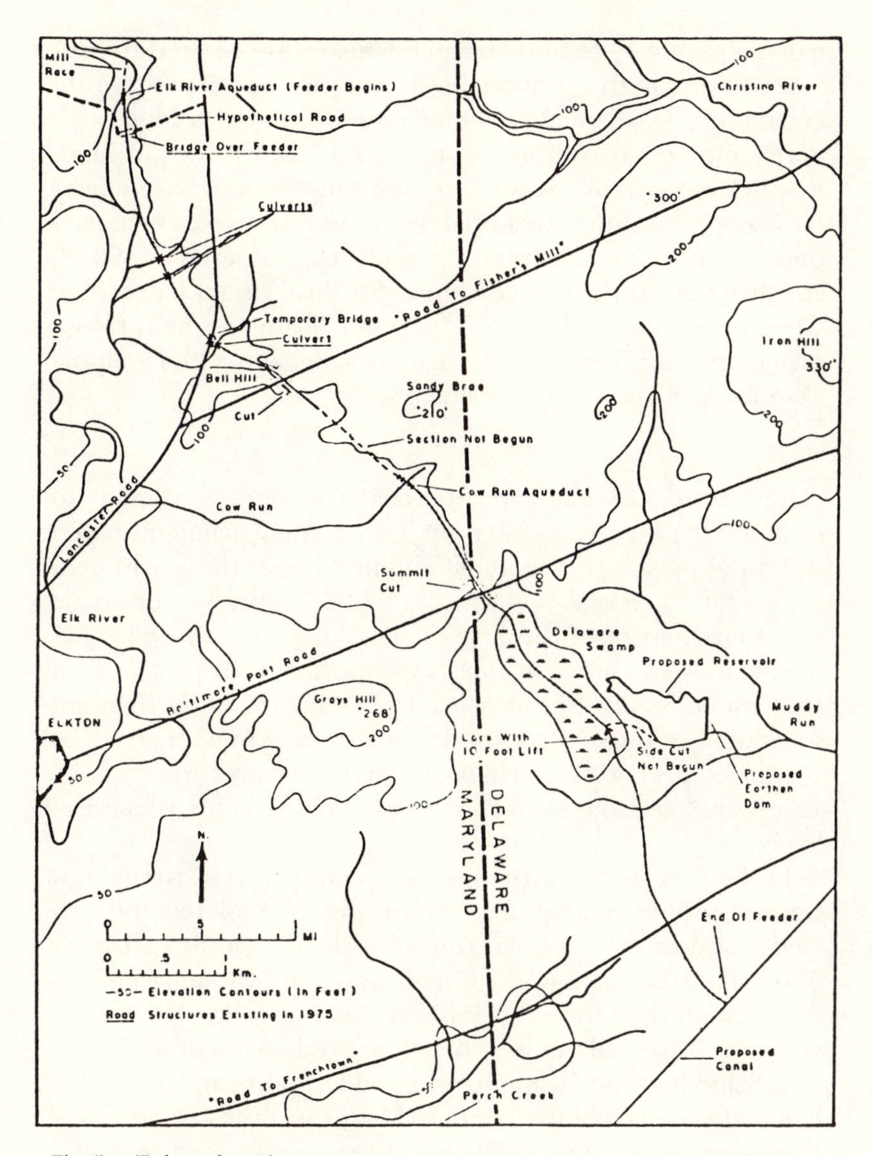

Fig. 5. (Enlarged area).

In the summer of 1803 Latrobe became one of the company's two chief surveyors, and in January 1804 the company appointed him their engineer. The directors chose the northernmost of the canal routes examined in 1769, from the Elk River at the head of the Chesapeake to the Christiana River near Wilmington on the Delaware, but Latrobe actually began work on a five-mile branch canal which was required to provide water for the summit level. For two years Latrobe directed the work, which included construction of bridges, culverts, and an excavation through a hill. But in the end, in a manner remarkably parallel to the Schuylkill and Susquehanna Company, many stockholders refused to make their payments, the company's treasury went dry, and late in 1805 all construction stopped.[84] Not only did this bring a halt to the Chesapeake and Delaware Canal, but it was the final spasm of activity in a fifteen-year drive to implement the Philadelphia plan for internal improvements.

* * *

The results of this late eighteenth-century movement to establish in America a system of internal improvements based upon the most advanced civil engineering in the world generally have been labeled a failure.[85] Certainly the dramatic economic impact of the Erie Canal, completed in 1825, has overshadowed earlier attempts to improve transportation in the United States. But measured by standards other than immediate economic effect the Philadelphia plan appears to be the major precursor, perhaps the necessary precursor, of the Erie Canal and the era of internal improvements which followed.

In the first place, of the improvement projects carried out under the Philadelphia plan, three were completed and carried considerable traffic: the Philadelphia and Lancaster Turnpike, the Conewago Canal, and the Susquehanna improvement. The three unfinished canal projects were in the short run financial failures, but they lived on as corporations. The Schuylkill and Susquehanna, and Delaware and Schuylkill companies were combined in 1811 to form the Union Canal Company. After 1815 that company was reorganized and refinanced, and finally constructed a working canal between the Schuylkill and Susquehanna rivers which operated for

[84] My remarks on Latrobe and the Chesapeake and Delaware Canal are brief because I have published the story elsewhere: Stapleton, *Engineering Drawings,* 12–17; Darwin H. Stapleton and Thomas C. Guider, "The Transfer and Diffusion of British Technology: Benjamin Henry Latrobe and the Chesapeake and Delaware Canal," *Delaware History* 16 (Fall–Winter 1976): 127–38.

[85] E.g., Taylor, *The Transportation Revolution,* 32.

nearly sixty years. The Chesapeake and Delaware Canal Company was revived somewhat later than the Union Canal but, after it decided on a new route farther down the Delmarva Peninsula, completed its canal in 1829. That canal still functions as a crucial link for East Coast marine traffic.[86] All things considered then, the visions of the eighteenth-century promoters were valid. What they did not complete was accomplished by their successors.

Second, the substantial efforts of 1791 to 1805 had created a pool of knowledge and inspiration about internal improvements which was drawn on by Philadelphians and others in succeeding years. Their works had a significant "demonstration effect." The Philadelphia and Lancaster Turnpike has been credited with initiating a turnpike boom throughout the Middle Atlantic states and New England.[87] The canals, finished or unfinished, were examined by a number of technically inclined persons over the years as physical manifestations of canal technology.[88] In the cases of both the Union and Chesapeake and Delaware canals, one of the earliest acts of the directors of the revived companies was to visit and admire the works erected by Weston and Latrobe.[89]

But more important than the works themselves was the development of an appreciation of the skills and abilities requisite to the practice of civil engineering. Both Weston and Latrobe claimed that the major value of an engineer was his technical experience, which meant that the work would be carried out with economy and a minimum of error. They also

[86] Livingood, *Philadelphia-Baltimore Trade Rivalry,* 90–97, 105–13; Ralph D. Gray, "The Early History of the Chesapeake and Delaware Canal: Completion of the Canal (Conclusion)," *Delaware History* 9 (April 1960): 66–98; Theodore E. Klein, *The Canals of Pennsylvania and the System of Internal Improvement,* Classics Reprint Series, no. 1 (Bethlehem, Penna.: Canal Press, 1973), 11–13.

[87] Taylor, *The Transportation Revolution,* 17–18; Durrenberger, *Turnpikes,* 46–47.

[88] E.g., Henry D. Gilpin to Joshua Gilpin, 28 May 1823, Gilpin Papers, Historical Society of Delaware, Wilmington; Barbara Benson, ed., *Benjamin Henry Latrobe and Moncure Robinson: The Engineer as Agent of Technological Transfer* (Greenville, Del.: Eleutherian Mills Historical Library, 1975), 66–67; sketchbook V, nos. 23–24, *Microfiche Ed.* The Erie Canal commissioners visited the Middlesex Canal and the Susquehanna Canal, and perhaps others: Shaw, *Erie Water West,* 69; "Autobiography of John Davis, 1770–1864," *Maryland Historical Magazine* 30 (1935): 29.

[89] Minutes of the Committee of Survey, 6 May 1822, Chesapeake and Delaware Canal Papers, Historical Society of Delaware; *Report of the President & Managers of the Union Canal Company of Pennsylvania; to the Stockholders* (Philadelphia: John Bioren, 1818), 14.

assured the investors of what Weston called their "*zeal* and *industry* in promoting the object of their undertaking."[90] Time after time Weston and Latrobe satisfactorily demonstrated these virtues of technical competence and commitment to the success of the enterprises, so that company managers used superlatives in describing their performance. William Smith praised Weston's "great abilities, activity and experience in all the branches of his department, [so that he] . . . merited and obtained the perfect esteem of the Managers." The president and directors of the Chesapeake and Delaware Canal stated that Latrobe's work was "done in a manner which will bear the strictest scrutiny or comparison with others of the kind in America or Europe."[91] The two engineers provided indelible examples of the value of professional direction of engineering enterprises.

If the promoters learned about professional engineering by observing the results, numerous craftsmen, contractors, and other skilled persons obtained experience more directly by executing Weston's and Latrobe's designs and instructions. The best entry into the realm of these skilled workers is through Latrobe's papers. They show that he was concerned to find people competent to respond to his demanding standards, and he often found that men who had previously worked for Weston satisfied his criteria. But he also trained men to his standards, sometimes employing them on a series of works in order to be assured of retaining their skills. Weston and Latrobe thus helped build up a corps of people responsive to and knowledgeable of the best civil engineering practice. Of those whose names we know, many contributed to other internal improvement projects in the growing republic. Latrobe also trained two future canal and railroad engineers, William Strickland and Robert Mills.[92]

Ultimately, the significance of Weston, Latrobe, and the

[90] William Weston to Richard Peters, 4 May 1803, Society Miscellaneous Collection, HSP; Smith, *Historical Account,* 53–54; Stapleton, ed., *Engineering Drawings,* 69–70; Benjamin Henry Latrobe to Elias B. Caldwell, 17 January [1810], Letterbooks, *Microfiche Ed.*

[91] Smith, *Historical Account,* 68; *Second General Report of the President and Directors of the Chesapeake and Delaware Canal Company, June 3, 1805* [Philadelphia: 1805], 7.

[92] Stapleton, ed., *Engineering Drawings,* 12–14, 20, 24, 63–64, 68–69; s.v., "Robert Mills," *DAB*; Darwin H. Stapleton, "The Origin of American Railroad Technology, 1825–1840," *Railroad History* 139 (Autumn 1978): 72.

Philadelphia plan for internal improvements may be considered in light of its fundamental relationship to the planning of the Erie Canal. Few have appreciated how much the Erie Canal projectors were influenced by the Philadelphia plan. Several had backgrounds which indicate that they were well acquainted with the plan and its vision of controlling the commerce of the interior for the benefit of a seaport.

Gouverneur Morris, who was a close friend of Robert Morris, returned from Europe in 1798 a confirmed advocate of canals. He had in his protection while in London Robert Morris's son, who was a witness to the contract Colquhoun had drawn with Weston.[93] The New York merchant Thomas Eddy, another early promoter, was from a Philadelphia family.[94] Perhaps the earliest Erie Canal promoter was Elkanah Watson, who was in correspondence with Robert Morris, and who, as an ally of Philip Schuyler in creating the New York Western Inland Navigation Company, knew Weston's work well.[95]

When in 1810 the first commissioners were appointed to direct surveys of the Erie Canal route, they included Gouverneur Morris and Thomas Eddy. Eddy and another commissioner, DeWitt Clinton, who later staked his political career on the canal, argued that Latrobe should be hired to direct the surveys, but to no avail.[96] The next year the state legislature appointed a new commission to begin planning for construction. Robert Fulton was chosen as a commissioner this time.[97] He had taken an interest in canals while in Britain in the 1790s, briefly worked as a canal contractor, and wrote a book about canal technology.[98] A subcommittee of Eddy and Fulton, who was already acquainted with Latrobe, asked him to visit upstate New York to do an estimate of construc-

[93] Anne Cary Morris, ed., *The Diary and Letters of Gouverneur Morris,* 2 vols. (New York: Charles Scribners' Sons, 1888), 2: 128–29; *Facts and Observations in Relation to the Origin and Completion of the Erie Canal* (New York: N. B. Holmes, 1825), 6–9; Oberholtzer, *Robert Morris,* 278–80; S and S mins., 14 January 1793.

[94] Shaw, *Erie Water West,* 16, 21; William W. Campbell, ed., *The Life and Writings of DeWitt Clinton* (New York: Baker & Scribner, 1849), 27–29; s.v., "Thomas Eddy," *DAB.*

[95] *Facts and Observations,* 5–6; Watson, *Life and Times,* 314–18, 336–40, 357–69.

[96] Campbell, *Life and Writings,* 27–29.

[97] Shaw, *Erie Water West,* 45–46.

[98] John F. Bell, "Robert Fulton and the Pennsylvania Canals," *Pennsylvania History* 9 (July 1942): 191–96; Robert Harris, *Canals and Their Architecture* (New York: Frederick Praeger, 1969), 90.

tion. Latrobe showed interest, but his commitments kept him from accepting immediately. When he finally was ready, the commissioners decided his fee was too high, and they had some surveys done by Benjamin Wright, who had worked under Weston on the Western Inland Navigation.[99]

They then turned to William Weston, and despite a war with Britain, asked him to oversee the canal.[100] He declined the invitation, but he had already been sent the surveys done to that point and he advised the commissioners that it would be a difficult, but not impossible work. Moreover, he praised their ambition to build a canal for forecasting that

> should your noble plan, of uniting lake Erie with the Hudson, be carried into effect, you have to fear no rivalry. The commerce of the enormous extent of the country bordering on the upper lakes, is yours forever; and to such an incalculable amount as would baffle all conjecture to conceive.[101]

The commissioners proudly quoted these words in their reports of 1812 and 1814 as the assessment of a celebrated and eminent civil engineer. As late as 1816 the commissioners relied on Weston's survey of the Mohawk Valley for a preliminary determination of the Erie Canal's route.[102]

Yet that same year, just a year before Latrobe called for importing more civil engineers, the Erie Canal commissioners decided not to try again to hire a European engineer because, as they put it, "there is every inducement for preferring our own countrymen if the requisite scientific and practical knowledge can be found."[103] The specific grounds for this decision are unknown, but I suggest that the commissioners recognized that by 1816 an American engineering style was emerging, one which in historical retrospect appears to have been better adapted to the American landscape and society than English engineering.[104]

Both Latrobe's and Weston's works were criticized by some

[99] Stapleton, ed., *Engineering Drawings,* 23; Calhoun, *American Civil Engineer,* 25.

[100] Kirby, "William Weston," 120; Benjamin Henry Latrobe to James Madison, 8 April 1816, Letterbooks, *Microfiche Ed.*

[101] *Laws of the State of New York, in Relation to the Erie and Champlain Canals . . . ,* 2 vols. (Albany, N.Y.: E. and E. Horsford, 1825), 1: 75.

[102] Ibid., 1: 74–75, 81–82, 105–106; Shaw, *Erie Water West,* 64.

[103] *Laws of the State of New York,* 1: 117.

[104] Calhoun, *American Civil Engineer,* 24–50.

as too expensive, primarily because they valued durability and adhered to rigid standards of design and construction. Both insisted on masonry construction for bridges, locks, and aqueducts, for example. The Erie Canal commissioners probably recognized that the 360-mile canal they had in mind required a more flexible view of materials usage, and, indeed, when the canal was built it used wood heavily where English engineering tradition would have required stone.[105]

One suspects that the commissioners thought carefully about their need to appoint a person versed in the canal engineering skills possessed by Weston and Latrobe, yet who had an American perspective on problem-solving. They found their man in Benjamin Wright, that upstate New York surveyor who had worked under Weston. Certainly Wright was not acquainted with the full range of European canal technology, but he was soon ably assisted by Canvass White, who went to England to study canals in 1817 and returned to direct the construction of locks and aqueducts.[106]

In sum, the working out of the Philadelphia plan for internal improvements was a crucial step in initiating the development and appreciation of civil engineering skills in the United States, and in laying the groundwork for the nineteenth-century transportation revolution symbolized by the Erie Canal. From the pre-Revolutionary era, Philadelphians sustained a vision of promoting the economic growth of their city through engineering works, and eventually they employed two English-trained engineers to fulfill that vision. The specific objectives of the Philadelphia plan were not completed until the 1820s, but in the meantime the transfers of engineering technology embodied by Weston and Latrobe provided the foundation for the rise of an American engineering tradition.

[105] Kirby, "William Weston," 125; *Report of the President and Managers of the Union Canal Company of Pennsylvania* (Philadelphia: John Bioren, 1819), 8–9; Stapleton, ed., *Engineering Drawings,* 64–68.

[106] Calhoun, *American Civil Engineer,* 25, 27, 29, 104–106; s.v., "Canvass White," *DAB.*

III: Eleuthère Irénée du Pont: *Elève des Poudres* to American Gunpowder Manufacturer

Eleuthère Irénée du Pont transferred French gunpowder technology to the United States on the basis of his training and skill as a student, businessman, and technologist. The threads of his French education, his employment in the *Régie des Poudres,* the family decision which brought him to America, and his own exacting, inquisitive mind weave an important story in the history of technology.

Eleuthère Irénée du Pont was born in Paris on 24 June 1771, the second son of a government official. Pierre Samuel du Pont, his father, was well suited to the court of Louis XVI, being a witty intellectual who was able to talk politics, philosophy, economics, and science with nearly equal facility. He had many friendships in official ranks, and it was Turgot, the minister of finance, who suggested the names Eleuthère ("liberty") and Irénée ("peace") to Pierre for his son. It was in character, then, that Pierre had Irénée tutored extensively in his youth, and that he showed great concern for his son's scholarly development. The earliest letters which he wrote to Irénée included admonitions about his academic skills, and praise when he had done his work well.[1]

Key to Footnotes for Manuscript Collections at The Hagley Museum and Library, Greenville, Wilmington, Delaware

---

L1 = Group 1, Longwood Manuscripts (Pierre Samuel du Pont)
L2 = Group 2, Longwood Manuscripts (Victor Marie du Pont)
L3 = Group 3, Longwood Manuscripts (Eleuthère Irénée du Pont)
L5 = Group 5, Longwood Manuscripts (E. I. du Pont de Nemours and Co.)
W2 = Group 2, Winterthur Manuscripts (Pierre Samuel du Pont)
W4 = Group 4, Winterthur Manuscripts (Eleuthère Irénée du Pont)
Acc. 500 = Accession 500, Records of E. I. du Pont de Nemours and Co.: Lb 1 = Letterbook 1; Lb 2 = Letterbook 2
Acc. 501 = Accession 501, P. S. du Pont Office Collection
HML = The Hagley Museum and Library, Greenville, Wilmington, Delaware

---

[1] P. S. du Pont, 7 May 1782, to E. I. du Pont, L1; P. S. du Pont, 3 September 1782, to E. I. du Pont, L1; P. S. du Pont, 27 July 1784, to E. I. du Pont, W2.

What led Irénée and his father to decide that his formal training and future occupation should be with the *Régie des Poudres et Salpêtres* (Agency for Powder and Saltpeter) is unknown, although two factors may have been responsible. Science was intellectually popular in Paris during the 1780s and probably exercised a great attraction for any young scholar, and the *Régie* was one of the French institutions which had something to do with the sciences. Perhaps more important was the friendship of P. S. du Pont with Antoine Laurent Lavoisier,[2] who was then one of the officials at the head of the *Régie,* and a great figure in European chemistry. The elder du Pont clearly expected Irénée to benefit from Lavoisier's influence.[3]

Although Irénée may have been happy with his appointment to the *Régie* in 1788, there were moments two or three years later when he wished he was doing something else.[4] At one point his brother Victor admonished him severely about such thoughts:

> You speak of being employed as a naturalist and I believe that it is a folly. It would be possible as a result of your studies and the examinations you have had to be employed as an engineer or in many other positions in the military. For if, as it appears, our government becomes completely republican, the sciences will be abandoned, and a poor naturalist ten years from now will die of starvation.[5]

Even though he may not have liked it, he remained with the *Régie* for several years, from 1788 to 1792, leaving only when Lavoisier's forced resignation left him without a powerful patron.

[2] The personal relationship between P. S. du Pont and Antoine Laurent Lavoisier is apparent in the former's correspondence, a great deal of which is printed in R. Dujarric de la Rivière, *E.-I. du Pont de Nemours: Elève de Lavoisier* (Paris: Librairie des Champs-Elysées, 1954). This book also includes a large amount of information concerning the intellectual climate of Paris in the 1780s, but nothing to substantiate the title, which indicates that E. I. du Pont was a student of the great chemist.

[3] P. S. du Pont, 15 November 1789, to E. I. du Pont, L1; E. I. du Pont, 22 August 1791, to P. S. du Pont, L3.

[4] P. S. du Pont, 15 November 1789, to E. I. du Pont, L1; P. S. du Pont, 17 April 1790 to E. I. du Pont, L1.

[5] Victor du Pont, 10 October 1791, to E. I. du Pont (L2-41), L2. Irénée was an avid botanist throughout his life—see Norman B. Wilkinson, *E. I. du Pont, Botaniste* (Charlottesville: University of Virginia Press, 1972).

The *Régie des Poudres et Salpêtres* was organized in 1775 after a private contractor failed to supply the necessary amount of saltpeter (the basic ingredient in gunpowder) to the French government.[6] Turgot had asked Lavoisier to make a report on the state of the powder manufacture, and the great chemist suggested that the government take over the operation of most mills and refineries. That done, Lavoisier was appointed a *régisseur* (director), a position he held until 1791.[7] He is generally credited with a great improvement in the provision of gunpowder, because of his administrative ability and scientific skills. The *Régie* increased annual production, and the government profited from its monopoly on saltpeter production by selling the excess to private powder manufacturers.[8]

Lavoisier's chemical skills were utilized in numerous ways. As part of the reformation of the saltpeter industry, Turgot requested in 1775 that the *Académie des Sciences* conduct a competition for the best essay suggesting an improvement in saltpeter production, and appointed Lavoisier one of five commissioners to do the judging.[9] It took several years before the essays were collected and published under the title *Recueil des mémoires sur la formation et la fabrication du salpêtre* (Collection of Papers on the Formation & Manufacture of Saltpeter). Lavoisier included four papers of his own, and handled the job of editing them for publication. Throughout the rest of his life Lavoisier investigated the "mysteries of saltpeter,"[10] writing several more papers on the subject.[11]

The purity of saltpeter (potassium nitrate) was a major concern in powder manufacture because the impurities in the crude saltpeter were salts which could absorb moisture, and

[6] Saltpeter is the major constituent of gunpowder, the proportions being approximately three-fourths saltpeter, one-eighth sulfur, and one-eighth charcoal.

[7] Edouard Grimaux, ed., *Oeuvres de Lavoisier,* 6 vols. (Paris: Imprimérie Nationale, 1864–93), 5: 717; Douglas McKie, *Antoine Lavoisier* (Philadelphia: J. B. Lippincott Company, 1936), 46; Robert P. Multhauf, "The French Crash Program for Saltpeter Production, 1776–94," *Technology and Culture* 12 (April 1971): 164, 167.

[8] McKie, *Antoine Lavoisier,* 46, 52; Lucien Scheler, "Lavoisier et la régie des poudres," *Revue d'histoire des sciences et de leur application* 26 (1973): 193–222.

[9] Ibid., 46–47; Multhauf, "The French Crash Program," 167.

[10] McKie, *Antoine Lavoisier,* 48; Multhauf, "The French Crash Program," 167; J. R. Partington, *A History of Chemistry,* 4 vols. (London: Macmillan and Co., 1962), 3: 466–68.

[11] Grimaux, ed., *Oeuvres de Lavoisier,* 5: 391–745, contains Lavoisier's works on powder and saltpeter. See also Partington, *A History of Chemistry,* 3: 466.

thereby destroy the explosiveness of the powder. In the late eighteenth century all the world's natural saltpeter came from the Bengal region of India. In France, however, some was gathered by scraping stables, barnyards, and the walls of dwellings, and also collected from beds of saltpeterish earth formed by heaping animal waste in enclosed areas.

The refining of saltpeter began with mixing the scrapings, earth, or imported saltpeter with water and potash. Some impurities precipitated out of the mixture, leaving a clear liquid which was poured into a cauldron and boiled for several days. Glue was added to aid in bringing impurities (salts) to the surface where they were skimmed off. When the impurities no longer rose the liquid was allowed to cool, and as the saltpeter formed into crystals it was removed and placed in washing boxes. Cold water was poured over the crystals, dissolving and removing some additional salts. After the crystals were dried they were ready for manufacture into gunpowder.[12]

Lavoisier's chemical interest in saltpeter was undoubtedly aroused because France faced blockade by England in wartime, which shut off Indian supplies. The French had to make up the deficit through their artificial sources, one of which, scrapings, was not capable of much expansion. They therefore made a great effort to increase production through the establishment of artificial beds. Beds produced less saltpeter after refining than did Bengal crude saltpeter, however, and, among other things, the essay competition judged by Lavoisier was an attempt to find ways of increasing their productivity.

No matter what the source, there was always the problem of determining whether the refined saltpeter was as pure as necessary. During Lavoisier's tenure with the *Régie* one of his subordinates, Jean Riffault, devised a new test for purity which became the standard for years afterward. It involved comparing the density of a saturated solution of pure saltpeter with one of newly refined saltpeter. A difference of six to seven percent was allowed for evaporation and error, but it

[12] Oscar Guttmann, *The Manufacture of Explosives,* 2 vols. (London: Whittaker and Co., 1895), 1: 25, 32–33; [Auguste] Bottée and [Jean] Riffault, *Traité de l'art de fabriquer la poudre à canon,* 2 vols. (Paris: Leblanc, Imprimérie-Librairie, 1811), 1: 25–26, 41, 77–89.

was a sophisticated procedure which required a trained chemist and fine laboratory equipment.[13]

A great deal of Lavoisier's work was based on experiments which he carried out at the Arsenal refinery (the headquarters of the *Régie*) in Paris, and at his laboratory in the same building. The laboratory was extremely well equipped and Lavoisier had invested a large part of his own income in it.[14] While Lavoisier is less known for his own experimental work than for his genius in organizing information to substantiate his theories, he did a large amount of pioneering work, and his laboratory was known as a center of intellectual and chemical ferment.[15] One result of the chemical work was that the refining of saltpeter was rationalized and improved.[16]

Irénée's association with the *Régie des Poudres et Salpêtres* began about November 1787 when he was nominated to the position of *élève* (student) in that agency, with the provision that a year later he would have to take competitive entrance examinations in mathematics, chemistry, and physics.[17] His father believed that he would have to study mathematics and drawing (drafting) in order to prepare for the examinations, and had his tutoring continued.[18] Irénée was accepted by the *Régie* late in 1788 when he was ranked "first in the mathematics examination and did well in the chemistry and physics."[19]

The position of *élève* in the *Régie* dated back to 1779 when the King's Council of State decreed that:

> The *Régisseurs* will choose, for entrance into the employ of the *Régie,* some educated subjects of good reputation provided with

[13] Multhauf, "The French Crash Program," 179–80; Bottée and Riffault, *Traité de l'art,* 1: 101–102.

[14] Henry M. Leicester, *The Historical Background of Chemistry* (New York: Dover Publications, Inc., 1956), 139–40.

[15] Rivière, *E.-I. du Pont de Nemours,* 134–35, 137; Partington, *A History of Chemistry,* 3: 376–77.

[16] Multhauf, "The French Crash Program," 177, 181; Grimaux, ed., *Oeuvres de Lavoisier,* 5: 717.

[17] P. S. du Pont, 6 November 1787, to Victor du Pont, W2; P. S. du Pont, 20 December 1787, to Victor du Pont, W2; P. S. du Pont, 21 February 1789, to Victor du Pont, W2; Victor du Pont, 19 April 1788, to E. I. du Pont, L2.

[18] P. S. du Pont, 6 November 1787, to Victor du Pont, W2; P. S. du Pont, 7 May 1788, to Otto, W2. Irénée's notes show that in the intervening year he attended lectures in physics and botany: box 10, L3.

[19] P. S. du Pont, 21 February 1789, to Victor du Pont, W4; E. I. du Pont, 23 October 1788, to Victor du Pont, W4.

> the chemical and mechanical knowledge necessary to this discipline; they will name to fixed positions which become vacant only those who will have been admitted previously according to the statement furnished by them each year to the *Administration générale des finances*; . . . .[20]

Because there is no adequate history of the *Régie*, the manner in which this decree was implemented is unknown, but a law passed by the National Assembly in 1791 uses language reminiscent of both the original decree and the experience of E. I. du Pont.

> No one will be able to attain employment in the *régie des poudres & salpêtres,* without having been an *élève* . . . ; and in order to obtain a commission as *élève,* it is necessary to be at least 18 years old, and to submit to a competitive examination in geometry and elementary mechanics, experimental physics and chemistry.[21]

This appears to have been an affirmation of former policy. The law also stated that all vacancies in the *Régie* were to be filled with the best candidate from the rank below, with *élèves* forming the lowest rank, and the *régisseurs* the highest. There were four of the former and three of the latter at that date.[22]

From fragmentary evidence it is possible to make up an outline of Irénée's progress through the *Régie.* In 1789 and probably in 1790 he attended lectures in natural history, physics, and chemistry offered by important figures in French science. At the Collège Royale he heard Louis-Jean-Marie Daubenton lecture on plant taxonomy. (Prior to joining the *Régie,* in 1785, he had attended Daubenton's lectures on botany and mineralogy.) He saw J. A. C. Charles's popular lectures and demonstrations of experimental physics. He had at least one lecture by Jean d'Arcet on fossil remains and the classification of minerals. His instructor in chemistry was "M. David," who cannot be identified, but Irénée also used a manuscript copy of Antoine Fourcroy's textbook *Philosophie*

[20] *Arrêt du Conseil d'Etat du Roi, Portant Réglement pour l'exploitation pendant six années de la Régie des Poudres & Salpêtre. Du Septembre 1779* (Paris: Impriméric Royale, 1779), 3.

[21] *Loi Relative à la fabrication & vente des Poudres & Salpêtres . . . le 19 Octobre 1791* (Paris: Imprimérie Royale, 1791), 6.

[22] Ibid., 7–9.

chymique, which was not published until after his studies were completed.[23]

The importance which the administrators of the *Régie* attached to this scientific instruction is not clear, but one essay which Irénée wrote at this time indicates that he was attempting to apply it to the process of saltpeter refining. In "Observations on the Six Kinds of Salts Found in Plaster . . ."[24] which applied chemical theory to the difficult problem of removing salts from the scrapings used in France for saltpeter manufacture, he referred to several chemical works and the chemical instruction he had received as the basis for his inquiry and to the refinery at the Arsenal as the location of his observations.[25] He also wrote a somewhat less theoretical paper describing the process of saltpeter refining.[26]

In 1790, perhaps after the period of formal instruction was complete, Irénée spent several months at Essonne, an important powder works southeast of Paris.[27] There he undoubtedly familiarized himself with the various stages of powder manufacture, from mixing ingredients to packing and storing. The results of this visit were two extensive essays, "Memoir on Powder," and "Memoir on the Construction of Powder Mills."[28] These differ from his earlier essay, "Observations . . . ," since they are not based on chemical theory, but manifest a very practical and thorough knowledge of the processes, problems, and machinery involved. Irénée's education under the *Régie* appears to have been well balanced between academic work and a thorough training in the technical aspects of powder manufacture.

With this background he may have advanced in 1790 or

[23] S.v., "Louis-Jean-Marie Daubenton," "Jacques-Alexandre-César Charles," "Jean d'Arcet," "Antoine François de Fourcroy," in *Dictionary of Scientific Biography,* ed. Charles Coulston Gillespie, 14 vols. (New York: Scribners, 1970–76); L3-2227; L3-2296; L3-2285; L3-2292; L3-2295; L3-2331. At one point Irénée referred to Fourcroy as "mon maître" which was probably a figure of speech rather than an indication of an educational relationship (L3-2313). Fourcroy's textbook is L3-2312.

[24] L3-2313.

[25] A series of observations on the refining process, titled "Notes Taken at the Refinery of Paris," are in Acc. 37, HML.

[26] "Mémoire sur le salpêtre," Acc. 519, HML; also L3-2316.

[27] See Victor du Pont, n.d. [1790], to E. I. du Pont (L2-32), L2; P. S. du Pont, 17 April 1790, to E. I. du Pont, L1; and P. S. du Pont, 3 May 1790, to E. I. du Pont, L1.

[28] "Mémoire sur la poudre," Acc. 519, HML; "Mémoire sur la construction des moulins à poudre," Acc. 37, HML.

1791 from *élève* to the rank of *contrôleur* or *commissaire comptable,* having the duties of an accountant.[29] The most telling evidence for such an advance is that in 1791 Irénée was talking about receiving an appointment as an *inspecteur-général,* the fourth rank in the *Régie,* which would have been a rapid move up the ladder from *élève.*[30]

This outline conforms closely to the statement which Irénée's father made in 1787 about his future.

> In a year he will begin as *Elève les cours de chimie et d'histoire naturelle.* He will then be for two more years in my household. Then he will go to Essonne with six hundred francs, then he will succeed to positions according to his capacity. If he remains a subordinate the work will not be so good. If he is able to claim *Inspecteur générale,* his road will have the greatest outlook; it is, however, only a moderate income. . . .[31]

There is little support for the popular notion that Irénée received his training in France as a student of Lavoisier,[32] even though both Irénée and Pierre told Thomas Jefferson that he did.[33] Lavoisier had some overall responsibility as *régisseur* for the instruction of the *élèves,* and it is true that he did work with a number of young chemists, including Séguin, Buquet, Hassenfratz, and Adet. He may even have supplied funds for their researches.[34] But E. I. du Pont's education was not a personal tutorial by the great chemist: it was a prescribed training in the administrative structure of a government agency.

It is important to remember that Irénée's work as an *élève*

[29] "I would a hundred times rather see you employed in bookkeeping at the Arsenal than at the mill of Essonne, and you know well that you would have better chances of advancement." P. S. du Pont, 17 April 1790, to E. I. du Pont, L1. In 1791 the next rank above *élève* was listed at *commissaire comptable* in *Loi relative à la fabrication* (n. 21), but Bottée and Riffault (n. 12), cxx, referring to the same date call the next rank *contrôleur.*

[30] E. I. du Pont, 15 August 1791, to P. S. du Pont, L3; E. I. du Pont, 22 August 1791, to P. S. du Pont, L3; Bottée and Riffault, *Traité de l'art,* cxx.

[31] P. S. du Pont, 6 November 1787, to Victor du Pont, W2.

[32] E.g., William S. Dutton, *Du Pont: One Hundred and Forty Years* (New York: Charles Scribner's Sons, 1942), 11–12; Rivière, *E.-I. du Pont de Nemours,* 136.

[33] E. I. du Pont, 20 July 1803, to Thomas Jefferson, L3; P. S. du Pont, 17 December 1800, to Thomas Jefferson (W2-588), W2.

[34] Rivière, *E.-I. du Pont de Nemours,* 134, 136; McKie, *Antoine Lavoisier,* 41; Partington, *A History of Chemistry,* 3: 368; also see articles on the scientists named in the *Dictionary of Scientific Biography* and the *Dictionnaire de Biographie Française.*

was preparation for administration and not for being a powder worker. The model which Lavoisier provided was of an administrator intimately concerned with the processes involved in the manufacture of powder through his chemical research, but who did not take part in them. Undoubtedly that was the sort of work to which Irénée aspired.

Irénée's preparation for a career in the *Régie des Poudres et Salpêtres* was interrupted by the French Revolution, which began with the formation of the National Assembly in 1789. Many careers were disrupted, ended, or forever changed in the following decade of governmental confusion, foreign war, and domestic bloodshed. Lavoisier was forced to resign from the *Régie* in 1791 and guillotined during the "Terror" of 1794, because of his long tenure as an official of the private corporation which collected taxes for the government. Even membership in the assembly gave no immunity from trouble, as it was periodically purged by whatever faction was in control. On the other hand, the Revolution was an exciting time for many who found adventure in military expeditions, who found the Revolutionary government listening to ideas for political and economic change, and who were called to perform extraordinary services in times of crisis. Irénée was involved in the excitement more than many Frenchmen because he was a young intellectual, the son of a politician, and lived in Paris, the seat of government.

During the last year that Irénée was with the *Régie* there was considerable anxiety in the du Pont family concerning what offices were open to him. Some of this discussion was due to his marriage to Sophie Madeline Dalmas, whom his father did not believe he could support with his current income.[35] But the resignation of Lavoisier was a strong blow to his hopes for promotion, and in 1792 he joined his father in the printing house which the elder du Pont had recently organized.[36] There Irénée got the opportunity to exercise the administrative training which he had received, because his

[35] E. I. du Pont, 15 August 1791, to P. S. du Pont, L3; E. I. du Pont, 22 August 1791, to P. S. du Pont; P. S. du Pont, 26 August 1791, to E. I. du Pont, L1; P. S. du Pont, n.d. [1791], to Madame Dalmas, L1.

[36] E. I. du Pont, 3 December 1800, to Louis de Tousard, W4 (". . . I left [the *Régie*] in January, 1792 . . .").

father became politically suspect due to his defense of the king, was arrested, and later had to go into hiding at the du Pont country home for over a year. The son was left in charge of the new printing house and by all accounts did a very creditable job.[37] The experience which the young man acquired in business was invaluable to him when he went to America.

During this period with the printing house Irénée still had occasional opportunities to learn about developments in gunpowder manufacture. He maintained a relationship with Lavoisier, who had an interest in saltpeter refining even after he had left the *Régie des Poudres.*[38] Irénée had this opportunity due to his father's continued friendship with Lavoisier, and to the publication of some of his works by the du Ponts.

In January 1794, Irénée was chosen the commissioner for the manufacture of saltpeter in his section of Paris, and served as secretary of a committee on saltpeter manufacture at least until July of that year.[39] He was probably asked to write pamphlets on the establishment of artificial saltpeter beds as part of the frantic campaign to meet the army's soaring demand for powder. Citizens were expected to engage in saltpeter production as a patriotic duty, and numerous pamphlets and articles on the subject were published.[40] Perhaps Irénée also acted as an instructor and inspector among the people; but it is certain that he was called on as a man with special knowledge.

There is circumstantial evidence for Irénée's continued awareness of developments in gunpowder manufacture at this time in an essay he wrote after deciding to establish a powder works in the United States. Although the essay is not dated, it was probably written late in 1800, before he returned to France for equipment and advice. In it Irénée compared cur-

[37] Note the power of attorney, P. S. du Pont, 14 June 1793, to E. I. du Pont, Acc. 276, HML.

[38] McKie, *Antoine Lavoisier,* 48.

[39] E. I. du Pont, 23 January 1794, to Sophie du Pont, L3; E. I. du Pont, 25 January 1794, to Sophie du Pont, L3; E. I. du Pont, 30 January 1794, to Sophie du Pont, L3; E. I. du Pont, 15 July 1794, to Sophie du Pont, L3.

[40] Bottée and Riffault, *Traité de l'art,* cxxvii–cxxviii. There is a volume of pamphlets and broadsides in box 3, L1, dating from 1791 to 1793, most of which were published for the Arsenal district of Paris, and all of which concern the manufacture of saltpeter by citizens.

rent American methods with French methods and noted that recent French improvements had greatly increased powder-mill productivity. The confidence with which he expressed himself is that of a man who was constantly informed about powdermaking technology.[41]

But the printing business absorbed most of Irénée's energies during the eight years after leaving the *Régie.* The du Ponts' careful policies kept their business out of political trouble, and Pierre's wide contacts among intellectuals gave it a large volume of work. It was, however, their journal, *L'Historien,* an excellent periodical which expressed well thought-out, though critical opinions on political issues, that finally led to difficulty. Because they were responsible for its publication both Irénée and Pierre were arrested shortly after the *coup d'état* of the Directory on 5 September 1797.[42] But Pierre had enough friends in the new government to obtain their release from jail in two days. Pierre may have made an agreement with the authorities whereby he was to abstain from further involvement in politics in return for the dropping of charges.[43]

It was this experience which turned the elder du Pont to the United States as an outlet for his activities. His friendships with the American ambassadors—Franklin and Jefferson, especially—had given him a favorable opinion of the new republic. Pierre had already written a great deal about the country, although he had never been there.[44] His other son, Victor, had lived in the United Sates almost continuously from 1787 to 1797 as a member of the official French delegation. Victor became a great admirer of the American character and knew many Americans.

During the rest of 1797 and in 1798 plans formed in Pierre's mind for moving his family to America. He first sought official sanction for leaving France, and persuaded the National Institute to appoint him an "explorer" of North America for his section, *classe des sciences morales et politiques.* Irénée was to

[41] "Projet d'établissement d'une fabrique de Poudre de Guerre et de Chasse dans les Etats-Unis" (hereafter "Projet d'établissement . . ."), box 11, L3.

[42] Mack Thompson, "Causes and Circumstances of the Du Pont Family's Emigration," *French Historical Studies* 6 (Spring 1969): 60–61.

[43] Ibid., 62.

[44] Ibid., 62–64.

be his secretary.[45] Pierre's imagination was not, however, content with such a simple activity; he developed a plan for a *Compagnie d'Amérique* which would establish a colony on about 100,000 acres somewhere in Virginia, Kentucky, or Tennessee. The du Pont family would lay out the major settlement and control the industrial and commercial activities. Pierre was to be the director of the company, with Bureaux de Pusy (his second wife's son-in-law) and his two sons as associates. He expected Thomas Jefferson, then vice president, to be available as an adviser.[46]

Investors showed little interest in Pierre's scheme, however, and the need for capital led him to transform it into a plan for a great international commission house. It was to handle the business of European firms trading in the United States and was to be based in New York.[47] Even so, the amount of capital collected fell far short of what was needed. The family raised money by selling the printing house and its book shop, selling family lands, and mortgaging the family estate, Bois des Fosses. In the end only 455,000 francs were paid in for company stock out of a projected total of four million. The only land obtained was some essentially worthless acreage in Kentucky which an investor had exchanged for stock.[48]

It took some time before all was in readiness for the family to leave France. They spent 1798 and most of 1799 trying to interest investors, clearing up family affairs, and studying English. The family party which finally left France on 2 October 1799 included Pierre, his two sons and their families, and Irénée's brother-in-law, Charles Dalmas. Their ship, the *American Eagle,* had a long and difficult voyage and landed at Newport, Rhode Island, on 3 January 1800.[49]

The family soon traveled to New York City, then moved into "Goodstay," a comfortable residence across the Hudson at Bergen Point, New Jersey. A notice appeared in New York

[45] Ibid., 66; P. S. du Pont, 2 Vendémiare VI [1797], to [J.-A.-C.] Charles, folio 3, folder labeled "Unknown Source II," box 9, Group 8, Longwood Manuscripts, HML.

[46] Thompson, "Causes and Circumstances," 66–73.

[47] Ibid., 70.

[48] Norman B. Wilkinson, "The du Ponts Come to America, 1797–1802" (unpub. research report, Hagley Museum, July 1955), 7–8, 16.

[49] Ibid., 12–14; Thompson, "Causes and Circumstances," 76.

newspapers that "du Pont de Nemours, père, fils et cie.," with a central office in Paris and branches in New York and Alexandria, Virginia, was ready for business.[50] But this new business had little success, and the family was forced to seek new opportunities. The elder du Pont drew up a list of possible ventures late in 1800, the eighth of which was black powder manufacture.[51] Colonel Louis de Tousard, a French *émigré* and inspector of artillery in the United States Army, evidently gave the family the original idea for a black powder company, and aided Irénée in the search for a good millsite. For his help he was rewarded with a small share in the profits of the powder works.[52]

E. I. du Pont carefully assessed the state of powder manufacture in the United States before the family made its decision. He visited the Decatur and Lane powder works in Philadelphia in November of 1800, and found that their processes were inefficient and out of date.[53] He wrote afterward that:

> There already exist in the United States two or three mills which make very bad powder and which do, however, a very good business. We will give an idea of the ignorance which they have of their art from those which up to this time have attempted this branch of commerce, and we will take for example the factory which has the best reputation, the one which now supplies the government. Nine years ago the entrepreneurs of this factory, who are merchants of Philadelphia, sent for a worker from Batavia, who makes powder for them as he saw it made in his country—as they made it perhaps fifty years ago in that Dutch province [i.e., the Netherlands].
>
> They use saltpeter from India which is infinitely better than that which is produced in France, but they refine it badly. They use it so impregnated with moisture and seawater that it is impossible, when they follow other good processes for manufacturing their powder, to have it half as strong as that which would be made with very dry and pure saltpeter.[54]

[50] Wilkinson, "The du Ponts," 14.
[51] Ibid., 17–18.
[52] Gabrielle Josephine du Pont, "Nôtre Transplantation en Amérique" (W3-5616) W3; "Acte d'Association," box 49, L5.
[53] Entry for 20 November 1800, "Dépense Générale," box 15, L5.
[54] "Projet d'établissement," box 11, L3.

This view of the technological backwardness of American powder manufacture is contradicted by Van Gelder and Schlatter in their *History of the Explosives Industry in America.* Referring to the period before 1800 they state that:

> Technically colonial [sic] mills were abreast of the English and probably in advance of them, as they received improvements from the continent before England and the shortage of labor provided an incentive for the exercise of the colonists' ingenuity. This was particularly true of power which was used more and to better advantage than in England.[55]

Such a comparison is puzzling as the English were particularly innovative in this period, introducing presses for caking the powder, destructive distillation of charcoal, and steam drying of the powder.[56] The Royal Gunpowder Factory at Faversham was a model of technical excellence between 1783 and 1815.[57] American methods could not possibly have been superior to the English and yet have merited such an abysmal rating from E. I. du Pont.

Unfortunately, there is no history of the early American black powder industry which might resolve these differences. The origins of the industry lie in the outbreak of the Revolutionary War when the country was cut off from its normal British supplies, and did not yet have a dependable ally. As part of a general campaign for self-sufficiency, government-encouraged powder mills were established in a number of places. One Pennsylvania report (of 1776) mentioned four mills under construction, one a large works with a saltpeter refinery, a pulverizing mill, a graining mill, and two stamp mills.[58] Benjamin Franklin prevailed upon Lavoisier to send two powdermen from the *Régie des Poudres* to the United States, and they spent 1777–79 traveling throughout the Mid-

[55] Arthur Pine Van Gelder and Hugo Schlatter, *History of the Explosives Industry in America* (New York: Columbia University Press, 1927), 69–70. Although discussing the period up to 1800, Van Gelder and Schlatter use the term "colonial."

[56] Guttmann, *The Manufacture of Explosives,* 1: 68, 204, 225.

[57] Arthur Percival, "The Faversham Gunpowder Industry," *Industrial Archeology* 5 (1968): 13–15.

[58] *Pennsylvania Archives,* first series (Philadelphia: Severns, 1853), 4: 765–67. See also the excellent article by David L. Salay, "The Production of Gunpowder in Pennsylvania During the American Revolution," *Pennsylvania Magazine of History and Biography* 99 (October 1975): 422–42.

dle Atlantic and New England states giving instruction.[59] The quality of the first American powder was probably very poor,[60] although it may have improved.

The impetus given by the Revolutionary War was enough to allow powder manufacture to continue in the United States afterward. Drawing upon statistics for Pennsylvania, the most active industrial state and the one which had the large mill described by Irénée, there is no doubt that powder manufacture was well established by the date of his arrival in the United States. In 1790 the magazine *American Museum* reported twenty-one powder mills in Pennsylvania, and a year later listed seventeen by county.[61] Tench Coxe, in his *View of the United States* (1794) stated that: "Twenty-one powder mills have been erected in Pennsylvania alone since the year 1768 or 1770—much the greater part of them since the commencement of the revolution [sic] war; four new ones are now building in that state. . . ."[62] These figures are corroborated in a memorandum written by Peter Bauduy in 1801, which lists the locations of fifteen mills in Pennsylvania.[63] The *Statement of the Arts and Manufactures of the United States* (1810), based on data gathered by Tench Coxe, showed twenty-two Pennsylvania mills producing about a fifth of the total amount of American powder.[64]

The Pennsylvania mills were part of an American industry which exported a yearly average of fifteen thousand pounds of gunpowder between 1791 and 1810.[65] Irénée's total of "two or three mills" can hardly be taken as the literal truth. The next five words of Irénée's sentence—"which make very

[59] Denis I. Duveen and Herbert S. Klickstein, "Benjamin Franklin (1706–1790) and Antoine Laurent Lavoisier (1743–1794). Part II. Joint Investigations," *Annals of Science* 11 (December 1955): 275–76.

[60] Waldo G. Leland, *Guide to Materials for American History in the Libraries and Archives of Paris* (Washington, D.C.: Carnegie Institution, 1932), item 9409.

[61] *The American Museum,* or *Universal Asylum* (microfilmed: American Periodical Series, Eighteenth Century, reels 4–5), 7 (1790): 289; 9 (1791): 68.

[62] Tench Coxe, *A View of the United States of America* (Philadelphia: William Hall, 1794; New York: Augustus M. Kelley, 1965), 148.

[63] "Peter Bauduy, list of Mills," box 52, L5.

[64] Tench Coxe, *A Statement of the Arts and Manufactures of the United States of America (1810)* (Philadelphia: A Cornman, Jr., 1814; New York: Luther Cornwall Company, n.d.), 33.

[65] Timothy Pitkin, *A Statistical View of the Commerce of the United States of America* (New York: James Eastburn and Co., 1817; New York: Augustus M. Kelley, 1967), 59–61.

bad powder"—represent perhaps a more reliable evaluation of the American powder industry. In addition to the poor saltpeter refining technique, he also noted that the Frankford mill granulated the powder badly.[66] Since he knew that this was a major powder works and a government contractor,[67] other smaller mills probably passed beneath his notice.

One may speculate that the quality controls which existed in France were probably unknown in the United States. Trained chemists were not available, and whatever testing methods did exist must have been rough. These conditions—poor technique, small size, and lack of quality control—are probably what led the young Frenchman to make such a disparaging assessment of American black powder manufacture.

Soon after the family decision to establish a powder works was made, Irénée arrived at another conclusion: he must return to France for equipment, advice, and funds. Additional funding was important: although the family company, du Pont de Nemours, père, fils et cie., was eventually to take two-thirds of the shares of the new company,[68] Irénée knew that he had "barely . . . the funds necessary for construction."[69] In France he expected to find willing investors.[70]

His technological expectations he revealed in a letter to Louis de Tousard.

> The usefulness of this trip for our business will be to enable me to follow in detail the very great and useful improvements which have taken place in the manufacture of powder since I left that field in January, 1792, and in addition to procure various articles which would be very expensive or difficult to procure here, such as the bronze mortars to equip the stamp mill, the parchments for the graining mills, and the punches for repairing them, copper sheets for the pounding, &c. . . .[71]

Irénée decided to take the packet *Benjamin Franklin* to France,

[66] "Projet d'établissement," box 111, L3.

[67] "Notes from the Contract with Lane & Decatur dated 14 March 1800," file 75, box 8, Acc. 501, HML.

[68] "Acte d'Association pour l'Etablissement," box 49, L5; original in file 21, Acc. 146, HML.

[69] "Projet d'établissement," box 11, L3.

[70] P. S. du Pont, 1 December 1800, to J. A. A. Bidermann, L1.

[71] Ibid.; E. I. du Pont, 3 December 1800, to Louis de Tousard, W4; P. S. du Pont, 17 December 1800, to Thomas Jefferson (W2-588), W2.

leaving early in 1801, and returning after a few months there.[72] He left on 5 January 1801, and had an excellent 28-day passage to Le Havre. He was accompanied by his brother Victor who was to obtain business for the parent firm.[73]

In a short time Irénée was in Paris to consult with the administrators of the *Administration des Poudres et Salpêtres.* This had been the title of the *Régie des Poudres* since 1797, and its headquarters were still at the Arsenal. Irénée received a warm welcome from Auguste Bottée and Jean Riffault, the chief administrators, and found them willing to accede to his various requests for help.[74] He also traveled to Essonne, where he conferred with *commissaire* P.-M.-C. Robin about purchasing equipment.[75] The powder works at Essonne were important in the Revolutionary era, and had a large force of workmen to repair and maintain the works, manufacture utensils, and make barrels for packing and shipping the powder.[76]

Irénée had to meet with these workmen to order the equipment which he desired, and to learn anew the techniques which were in use, and he must have discussed with them the possibility of emigration to America. A master powder worker named Lecompte was seriously interested, as well as a young cooper, but after several months' deliberation neither decided to come.[77] Irénée also met with workmen at the Arsenal refinery,[78] probably for the same purposes. The equipment which Irénée ordered was essential to every phase of black powder manufacture—boilers, pans and utensils for refining saltpeter, and laboratory equipment to monitor the process. Copper sheets and forms for the stamp mills, cloth for bolting ground charcoal and sulfur, parchment for use as sieves in graining the powder, punches for repairing the parchment, two presses for caking powder, and two dozen copper utensils were or-

[72] Ibid.

[73] Receipt for passage on the *Benjamin Franklin,* 31 December 1800, box 7, "1800" folder, L3; E. I. du Pont, 5 January 1801, to Sophie du Pont, W4; E. I. du Pont, 3 February 1801, to Sophie du Pont, W4; Victor Marie du Pont, *Journey to France and Spain,* ed. Charles W. David (Port Washington, NY: Kennikat, 1971), 4.

[74] E. I. du Pont, [May 1801], to [Auguste] Bottée, L3.

[75] P.-M.-C. Robin, 24 June 1801, to E. I. du Pont, L3.

[76] Rivière, *E.-I. du Pont de Nemours,* 139–47.

[77] E. I. du Pont, 25 January 1802, to P.-M.-C. Robin, L3; [Auguste] Bottée, 15 April 1802, to E. I. du Pont, L3; P.-M.-C. Robin, to E. I. du Pont, L3; [Jean] Riffault, 15 May 1802, to E. I. du Pont, L3.

[78] Entry for "drinks for workers at the refinery," "1801" folder, box 7, L3.

dered from the *Administration.* From Marseille Irénée ordered horsehair canvas which was specially needed for bolting and for sun-drying the powder, and an "*éprouvette de Regnier.*"[79] Irénée may have intended to use the French powder as a standard against which to measure his own.

While he was yet in France, most if not all of the equipment was packed for shipping, and most accompanied him home. Some was retained and sent later due to the British embargo on French shipping, and it finally arrived on the *William and Mary* nearly a year later.[80] In addition to the equipment, Irénée had made at Essonne "copies of plans and designs" which he probably intended to use as a guide during the construction of his powder works. He also obtained a copy of a *mémoire et instruction sur les poudres,* an official document of the *Administration.*[81] The significance of what Irénée acquired in France was clear to him. As he told Jean Riffault less than a year later, "I have the pleasure to regard [my enterprise] as a colony of your *Administration.*"[82] Such a characterization was clearly appropriate, as Irénée not only received his early training with the *Régie,* but had now gotten advice, equipment, and plans from its successor, the *Administration.*

Irénée completed his business in Paris by officially organizing his new enterprise as E. I. du Pont de Nemours and Co., a *société* under the laws of France. The major provisions of the *Acte d'Association* included the setting of the capital invested at $36,000, divided into eighteen shares. Du Pont de Nemours, père, fils et cie., took twelve shares, and French and American investors took the rest. Irénée was appointed full-time director of the company, and was instructed to finish construction of the works sometime in 1802, at which time production was to begin. In order to insure financial regularity Irénée was "to follow in his accounts the principles of accounting established in France by the *Administration des Poudres et Salpêtres.*" This agreement was dated 21 April 1801.[83] Irénée

[79] The record of the purchases and packing can be found in the "1801" folder, box 7, L3; and box 15, L5.

[80] "Dépense Générale," box 15, L5; "1801" folder, box 7, L3.

[81] The "plans and designs" and *mémoire* have not survived.

[82] E. I. du Pont, 25 January 1802, to [Jean] Riffault, L3. See also E. I. du Pont, 27 April 1807, to de Bussy, L3.

[83] "Acte d'Association," box 49, L5.

Fig. 6. Eleuthère Irénée du Pont (1771–1834). Courtesy of Hagley Museum and Library.

had made a commitment which determined the course of the rest of his life.

The return to the United States was a troubled one because although the *Franklin* left Le Havre on 1 May it was detained for a few weeks at Portsmouth by the British before it was allowed to depart. Irénée finally arrived at Philadelphia on 14 July, and immediately began what became a long and tiresome business—looking for a millsite.

Irénée had definite ideas of the kind of site he wanted. The water supply should be generous and the available fall of water adequate. The land should have a supply of timber on it to be used in construction, thus reducing expenses. A sawmill already built on the site would be ideal for preparing construction timber. Besides these qualities in the site itself, he wanted a location close to a seaport, and near enough to the new national capital to make communication with and transportation to it easy. This limited the search to the states of New Jersey, Pennsylvania, Delaware, Maryland, and Virginia.[84] Irénée originally felt that the states bordering on the Potomac River would be better because the shorter winter (compared to Pennsylvania and New Jersey) would result in a longer working season.[85]

With these *desiderata* in mind Irénée had made at least two excursions to examine millsites in the fall of 1800 before leaving for France. On one tour he visited Paterson, Morristown, and Dutch Valley in New Jersey; then he went to Easton, Bethlehem, Doylestown, Philadelphia, and Chester in Pennsylvania; and finally to Wilmington, Delaware.[86] On another journey he visited Maryland and Virginia.[87] Tousard was active in the search during this period, and before leaving Irénée asked him to continue to look for a suitable location.[88]

Upon his return from France Irénée took up the search with a new impetus: under the terms of the *Acte d'Association* he was to start production by sometime in the next year, and any delay in purchasing a millsite made that timetable less achievable. Yet it was nine months before a purchase was made. Irénée revisited some locations, but added a number of new ones to his list: Frenchtown, Union, Somerset, New Brunswick, Spotswood, and Little Falls in New Jersey, as well as Esopus in New York.[89] The concentration on sites in northern New Jersey was undoubtedly due to Irénée's residence at Bergen Point, and to his appreciation of the natural features of the area, evidently no small matter in Irénée's mind.[90] At one point he seems to have found a site on a branch of the Raritan which satisfied him, but the Brandywine opportunity proved the better.[91]

[84] "De l'emplacement et des constructions nécéssaires pour l'établissement d'une fabrique de poudre," box 11, L3.

[85] E. I. du Pont, 3 December 1800, to Louis de Tousard, W4.

[86] E. I. du Pont, 2 October 1800, to Sophie du Pont, W4.

[87] P. S. du Pont, 14 November 1800, to Thomas Jefferson, W4.

[88] E. I. du Pont, 3 December 1800, to Louis de Tousard, W4.

[89] "Dépense générale," box 15, L5; "1802" folder, box 7, L3.

[90] E. I. du Pont, 20 September 1801, to Sophie du Pont, W4.

[91] E. I. du Pont, 10 February 1802, to Peter Bauduy, L3.

The possibility of purchasing an operating powder mill also appealed to the young Frenchman, as it would have had the advantages of possessing a millsite, buildings, and a workforce. He would have to rebuild the works partially and retrain the workers, but would be spared a great deal of trouble. He seriously investigated the Decatur and Lane works in Frankford, near Philadelphia, for that purpose, and evidently made an offer to purchase which was accepted by only one of the partners (Lane). Unable to consummate the deal, he nonetheless used the bargaining to learn about the state of American powder technology and the American powder market.[92]

For a number of reasons Irénée's attentions gradually settled on the Brandywine River near Wilmington, Delaware. There he found a colony of Frenchmen who were happy to help him in his search and to help him deal with sometimes difficult American laws and customs. Peter Bauduy became a particularly good friend, and his presence led Irénée to consider the Brandywine as the most attractive site.[93] Negative emotion also played a part in centering his attentions: Irénée decided he did not want to live in Maryland or Virginia because

> the location, the land, and the people are worthless. The slavery of blacks, who are rather numerous, the poverty of the land, the vanity, the luxury, the extravagant spirit and the haughtiness of the so-called rich men who for the most part do not have much, have rendered that class of people, and the workers generally, as bad as we have found them good in our vicinity [northern New Jersey] until now.[94]

More practically, he believed that the Hudson and Passaic river valleys lacked good transportation to market.[95]

The Brandywine fit Irénée's original criteria for a good river as it had an excellent flow and good fall. In the land of Jacob Broom about four miles above Wilmington were found the remaining characteristics of a good millsite: proximity to good ports (Wilmington and New Castle), adequate space for

[92] E. I. du Pont, 25 December 1801, to Louis de Tousard, L3; E. I. du Pont, 25 January 1802, to [Jean] Riffault, L3.

[93] E. I. du Pont, 20 September 1801, to Peter Bauduy, L3, E. I. du Pont, 2 January 1802, to Peter Bauduy, L3; E. I. du Pont, 25 June 1802, to Peter Bauduy, L3.

[94] E. I. du Pont, 20 September 1801, to Sophie du Pont, W4.

[95] E. I. du Pont, 30 November 1801, to Peter Bauduy, L3.

economical grouping of buildings, standing timber, and a sawmill. Bauduy informed Irénée of the availability of that plot in October 1801, and got an enthusiastic response from him, although he was concerned by what he considered a high price for the land—$8,000.[96] From that time the Broom property was always on Irénée's mind, and the negotiations involving him, his Wilmington friends, and Broom were constant.[97] At one point Irénée became frustrated by the haggling, and considered buying a site on the Raritan, but late in March 1802, he wrote to Broom that he was ready to make the purchase.[98] Because of Delaware laws, William Hamon, a French refugee from Santo Domingo who had become an American citizen, took the deed in E. I. du Pont's behalf.[99]

It was three months before the du Pont family moved from Goodstay to their new home, which was to become Eleutherian Mills. In mid-July they sent some of their most valuable possessions, including hunting dogs, powdermaking equipment, and merino sheep, by schooner to Wilmington, and themselves set out overland.[100] Bypassing Philadelphia, which was in the grip of a yellow fever epidemic, they arrived at their destination on 18 July 1802.[101] Irénée's first reaction after a few days on the Brandywine was that he had moved to a very remote section of the country. He told his brother that:

> . . . if the beginnings of our establishments are not as difficult as they have been on the banks of the Ohio or the Wabash, it is close enough for us to have an idea of what they would have been in the most savage country. . . . We have not yet been able to procure a servant, but I hope we will have one soon. Sophie has seldom been more fatigued.[102]

[96] Peter Bauduy, 5 October 1801, to E. I. du Pont, L3; E. I. du Pont, 7 October 1801, to Peter Bauduy, L3.

[97] E. I. du Pont, 30 November 1801, to Peter Bauduy, L3; E. I. du Pont, 27 December 1801, to Peter Bauduy, L3; E. I. du Pont, 2 January 1802, to Peter Bauduy, L3.

[98] E. I. du Pont, 10 February 1802, to Peter Bauduy, L3; E. I. du Pont, 28 February 1802, to Louis de Tousard, L3; E. I. du Pont, 30 March 1802, to Jacob Broom, L3.

[99] The papers recording the purchase are in box 49, L5.

[100] E. I. du Pont, 10 July 1802, to William Hamon, L3; "Dépense Générale," box 15, L5: E. I. du Pont, 15 July 1802, to Charles Dalmas, W4.

[101] E. I. du Pont, 17 July 1802, to Victor du Pont, W4.

[102] E. I. du Pont, 21 July 1802, to Victor du Pont, W4; see also E. I. du Pont, 15 August 1802, to P. S. du Pont, L3.

Nevertheless, Irénée soon gathered a workforce and began construction of the powder works. He had already decided what buildings he wanted to construct and what their spatial relations would be.[103] A refinery was to be constructed with three sections: storage for crude saltpeter, a saltpeter refinery, and a drying room. The refinery was to be equipped with brick furnaces surmounted by large copper cauldrons or boilers (six or seven feet in diameter) for dissolving and removing the waste from saltpeter. It would also have twenty or more crystallizing basins for cooling the solution of refined saltpeter, and a number of wicker boxes for rinsing the crystals removed from the basins. The drying room for the purified saltpeter was to have many windows to promote circulation.

There were to be several buildings along the millrace on the river. The pulverizing mill (or wheel mill) was to have a water wheel with gearing connecting it to two vertical stone wheels running in a circle. All three powder ingredients—charcoal, sulfur, and saltpeter—would be crushed under the wheels. In the same building, and powered by the waterwheel, were to be two bolters for saltpeter, and one for each of the other ingredients. Bolting, or sifting, took place immediately after the pulverizing so as to insure that only small pieces were used for powder. Irénée also believed that this mill would be used for storage of sulfur and charcoal.

Near the pulverizing mill and the stamp mill, but not on the millrace, would be the composition house, where the three ingredients were weighed, measured, and mixed. The mixture would then be taken to the stamp mill to be pounded for several hours under vertical stamps. This mill was equipped with a waterwheel, gearing to transmit the power to an axle with projecting spokes, and vertical stamps which were raised by the spokes. After being raised, the copper-sheathed ends of the stamps fell into mortars containing the gunpowder mixture and enough water to keep it moist. The pounding action of the stamps thoroughly incorporated the three ingredients so that when detonated a complete chemical reaction would take place.

[103] "De l'emplacement des constructions nécéssaires pour l'établissement d'une fabrique de poudre" (L3-2365), box 11, L3; "Evaluation de ce que peuvent couter les constructions nécéssaires pour l'établissement d'une manufacture de poudre en Amérique," box 49, L5.

Irénée did not plan for a press house where the incorporated powder could be forced into hard cakes and then broken into fragments. He did, however, import presses from France.[104] Perhaps he intended that step to take place in the graining house, which was also to be located along the millrace. Graining involved taking the pressed and broken fragments of powder and pushing them through sieves of different sizes. Large sizes were used for cannon and mortar powder, and smaller ones for carbine and rifle powder. Irénée hoped to replace the usual manual graining process with a water-powered machine which he knew to be used in France.

Both the stamp mill and the graining mill were to be rather distant from the other buildings, and each other, because after the powder left the composition house it had to be considered dangerous. These two buildings, along with the glazing mill, were to be constructed in the style peculiar to French powder mills: three high strong walls, one low wall (or no wall) on the river, and a light roof slanting down toward the river. This shape directed explosions away from the rest of the powder works, and tended to reduce damage to the building itself.

After graining, that powder which had been made for high-quality sporting powder would be taken to the glazing house, which was along the millrace and powered by a waterwheel. The wheel would turn gearing connected to several sealed barrels turning on their long axes which tumbled the powder until rounded and smoothed. Those characteristics gave it the ability to pour and resist moisture better. After graining, or graining and glazing, all powder would go to the dryhouse, from which it could be removed on calm, sunny days to be dried on canvas spread over tables. In rainy or windy weather the powder could be dried inside the building with heat from a fireplace stoked from the outside. The dryhouse was to be distant from other structures.

The final processing step would take place in a building not too far from the dry house but still isolated. This building would house a bolter which would separate all the fine dust from the processed powder, leaving only well-formed grains. In the same structure the finished product would be packed

[104] Item for presses in "Estimate of necessary expenses for the establishment of powder manufacture," box 11, L3.

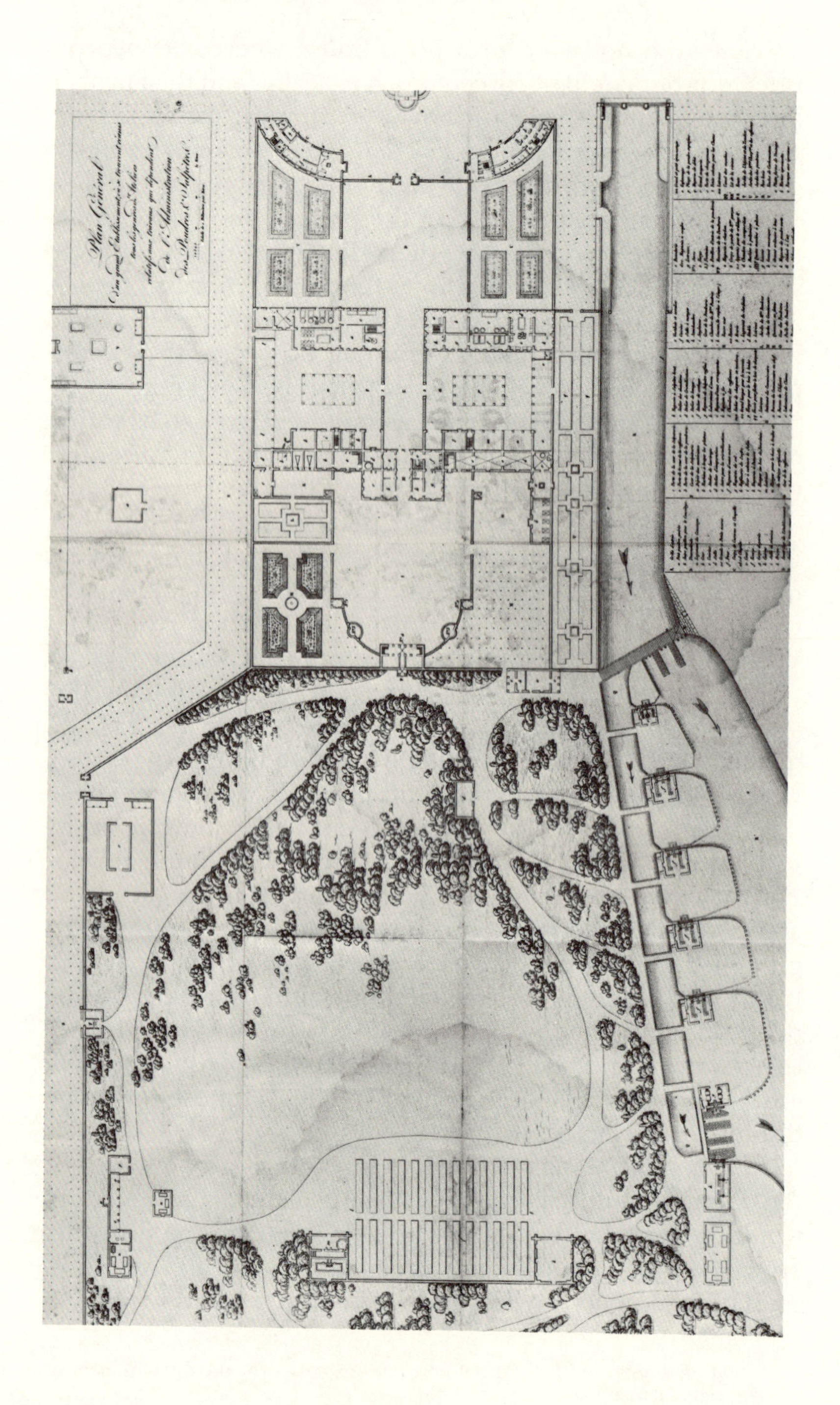

into kegs of standard sizes, mostly of twelve and a half and twenty-five pounds. From there the kegs would go to the magazine, the most distant structure of all. They were removed from the magazine only when shipped to market.[105]

The du Pont powder mills as finally constructed were sketched by Charles Dalmas in 1806.[106] The layout he shows does not fully conform to Irénée's ideal, especially in that the buildings are close to one another. That may have been the result of the narrowness of the Brandywine Valley, which limited the amount of level ground on which to build. Perhaps when Irénée had made his plans he was thinking in the same terms as Bottée and Riffault when they published a diagram of an ideal powder works in 1811 (fig. 7).[107] The spaciousness and garden-like surroundings they envisioned were not possible for Irénée, and Eleutherian Mills looks more like an industrial compound in the Dalmas sketch.

Nonetheless, the sketch (fig. 8) shows that Eleutherian Mills was fundamentally arranged in accord with the same principle of safety which Bottée and Riffault had in mind: that is, the buildings housing the dangerous processes of mixing, graining, and glazing, and the powder magazine were on the periphery. Other buildings, including the refinery, the compo-

[105] This outline of Irénée's intentions is drawn from his essay, "De l'emplacement et des constructions nécéssaires pour l'établissement d'une fabrique de poudre" (L3-2365) L3. To fill out the process of powder manufacture I have depended upon Bottée and Riffault, *Traité de l'art,* Guttmann, *The Manufacture of Explosives,* and process sheets from the files of Robert A. Howard, HML.

[106] Sketch displayed in the Hagley Museum, Greenville, Wilmington, Delaware.

[107] Bottée and Riffault, *Traité de l'art,* 2: plate 39.

←

Fig. 7. "General Plan of a Large Establishment, in which are gathered all kinds of Shops connected to the works which are dependencies of the *Administration des Poudres & Salpêtre,*" From Bottée and Riffault, *Recueil de planches relatives à l'art de fabriquer la poudre à canon* (1811), plate XXXIX. This is an ideal plan for a French powder works. The large, extensively divided building on the right contains shops to refine, test, and pulverize powder ingredients, and also houses the administrative offices of the enterprise. Along the river, powered by water from a race, are the stamp mills, glazing mill, and graining mill. On the left in the middle of the plan are the dry tables and dry house. At the top from left to right are the pack houses, gate, magazine, powder testing area, and charcoal kilns. Courtesy of Hagley Museum and Library.

Nº1
ELEUTHERIAN MILLS
1806.
By Chas. Dalmas.

sition mill, and miscellaneous outbuildings, were located in the central area. This layout also concentrated the workforce in the center, which probably permitted better oversight by the manager, since the refining and composition stages were relatively labor-intensive.

With his original plan in mind Irénée had now to select the building sites and begin work. He was the only person there who knew how a powder works functioned and what was required in construction, forcing him to draw up plans for his workmen at night and to direct their accomplishment during the day.[108] For nearly two years Irénée had on hand a force of workers which numbered from forty to sixty,[109] and which he was anxious to have do the work correctly. Three months after construction began he was complaining that:

> . . . I am obliged to supervise ceaselessly and to manage the work of a regiment of laborers and one troop of masons and two of carpenters, perpetually obliged to make myself understood to men who do not comprehend me, and who hardly comprehend their work any better, for it would be difficult to gather worse workers of any sort than what I am obliged to employ.[110]

Nine months later things were no better.

> I have had, since spring, more than forty workers on the *corps.* It is necessary that I plan the work with them; I lead them and

[108] E. I. du Pont, 29 October 1802, to Charles Dalmas, W4.

[109] Sophie du Pont, 15 September 1802, to Charles Dalmas, box 14, L3; E. I. du Pont, 12 June 1802, to P. S. du Pont, L3; E. I. du Pont, 3 August 1802, to Victor du Pont, L3; E. I. du Pont, 3 August 1802, to Victor du Pont, W4.

[110] E. I. du Pont, 18 October 1802, to Victor du Pont, W4.

←

Fig. 8. Eleutherian Mills, near Wilmington, Delaware, in 1806. At the top of the hill on the left are the du Pont residence and barn. Along the river on the left are the pulverizing mill (wheel mill), the stamp mill, the saw mill, and (further to the right) the glazing mill. They were powered by water from a race which began to the right of the dam in the Brandywine River, shown in the foreground. The row of buildings behind the mills and along the white fence begins on the left with the *L*-shaped refinery. The next two buildings, behind the glazing mill, are the composition house, and combination charcoal and dry house. To the right of the gate, and intersected by the fence, is the building usually identified as the home of the du Pont family before the Eleutherian Mills residence was completed. The blockhouse-like structure to the far right is the magazine where finished powder was stored. Other buildings are workers' cottages and outbuildings. Courtesy of Hagley Museum and Library.

supervise them. And in a foreign country, where one hardly knows the language, and where construction is also difficult, this process is truly frightening.[111]

These two statements suggest one reason why Irénée never achieved the goal which the *Acte d'Association* set of getting into production by 1802, or even his own later goal of the fall of 1803.[112] For some reason it was a long time before he was comfortable with the English language. Although he studied it before leaving France and at Goodstay,[113] when he came to the Brandywine he still could not conduct business in English. Much of the early company business was handled by Peter Bauduy, and most of Irénée's English letters were copies of translations by Bauduy.[114] Over five years after his arrival in the United States he admitted to his daughter that he could not yet write in English.[115]

That problem was compounded by Irénée's difficulties with his workforce. He complained that they were unable to carry out his orders without a thousand errors.[116] He believed that they were too easily frightened when the master mason got jaundice, since all the journeyman masons fled because they thought he had yellow fever.[117]

It is unlikely that Irénée had any problems with the Brandywine Valley millwrights whom he hired,[118] because they were among the most active and experienced in the United States. The most prominent American millwright of the period, Oliver Evans, came from their ranks.[119] They handled so much business that in 1797 thirteen of them issued a list of prices for standard items in mill construction.[120]

[111] E. I. du Pont, 12 June 1802, to P. S. du Pont, L3.

[112] E. I. du Pont, 7 February 1802, to P. S. du Pont, L3.

[113] E. I. du Pont, 18 November 1800, to Sophie du Pont, W4; E. I. du Pont, 2 November 1797, to Sophie du Pont, L3.

[114] This is apparent from the early letters and papers in L5. Bauduy was made an associate of the company in 1802—see "Articles d'agrément," file 21, Acc. 146, HML.

[115] E. I. du Pont, 23 September 1805, to Victorine E. du Pont, W4.

[116] E. I. du Pont, 29 October 1802, to Charles Dalmas, W4.

[117] E. I. du Pont, 5 November 1802, to Victor du Pont, W4.

[118] Accounts with millwrights, 1802–04, file 148, Acc. 146, HML.

[119] Grenville Bathe and Dorothy Bathe, *Oliver Evans* (Philadelphia: Historical Society of Pennsylvania, 1935).

[120] *Article of Agreement . . . for the rules and wages of millwrighting* [Wilmington, Del.: 1797]. Cf. Peter Bauduy, 20 August 1802, to E. I. du Pont, L5.

Although none of these millwrights could have been familiar with the exact structures Irénée wanted built, it is interesting to note that there were native analogues to the stamp and pulverizing mills of Irénée's powder works.[121] The stamp mill was probably the mill most necessary to the powdermaking process and may be thought the most peculiar to it, although similar machinery was in use in the United States for other purposes. The similarity of stamps in fulling mills is indicated by an offer of a New Jersey fuller during the Revolutionary War to convert his operation to a powder mill.[122] The correspondence was so close that an individual who made a map of Eleutherian Mills about 1820 labeled the stamp mill with the words "fulling mill—pounding."[123] There were a number of fulling mills in New Castle County (northern Delaware) at the time of E. I. du Pont's arrival.[124] Oil mills, which pressed linseed oil from flax seed, were also common in Pennsylvania and New Jersey during the same era. At least one in New Jersey used stamps, and in 1801 Peter Bauduy noted several Pennsylvania mills that had been converted from powder making to oil production.[125]

The pulverizing mill, or wheel mill, with its pair of large millstones on horizontal axles and running in a circle, also had its analogues in America. Many early oil mills were operated in that fashion.[126] Tanyards also used wheel mills with one or two stones to crush tanbark; most were powered by horses or mules, but a few used water power before 1800.[127]

[121] Of course, it is possible that some of them may have been familiar with powder mills in Pennsylvania, but the machinery which American powder mills used is even less known than that used by other mills.

[122] *Pennsylvania Archives,* 1st ser., 4: 709.

[123] Sketch of Eleutherian Mills, ca. 1815–25, Acc. 469.

[124] Carroll Wirth Pursell, Jr., "That Never Failing Stream" (M.A. thesis, University of Delaware, 1958), 20.

[125] Harry B. Weiss and Grace B. Weiss, *Forgotten Mills of Early New Jersey* (Trenton: New Jersey Agricultural Society, 1960), 12; Whitney Eastman, *The History of the Linseed Oil Industry in the United States* (Minneapolis: T. Dennison and Company, 1968), 25–26; "Peter Bauduy, list of mills," ca. 1801, box 52, L5.

[126] Greer Scheetz, "Flax Seed Mills," *Papers of the Bucks County* [Penna.] *Historical Society* 4 (1917): 725–26.

[127] J. Leander Bishop, *A History of American Manufactures,* 2 vols. (Philadelphia: Edward Young and Co., 1864), 1: 453–54; Harry B. Weiss and Grace M. Weiss, *Early Tanning and Currying in New Jersey* (Trenton, N.J.: 1959), 29–30. Cf. Peter C. Welsh, *Tanning in the United States to 1850* (Washington, D.C.: Smithsonian Institution, 1964), 8, 18.

New Castle County had a number of tanyards in this era.[128] Considering their probable experience with analogous structures, it seems likely that Irénée's millwrights had no difficulty building what he wanted.

His feelings about his workmen may have resulted from personal prejudices toward Irishmen, since most of his workforce was recently arrived from Ireland. At one time he wrote:

> It will be, I believe, impossible to educate a skilled worker from the race of Irish workers which they have in this country. I have employed nearly a hundred of them this last year and in this quantity there were no more than two whom I would want to work for me in the mill.[129]

He wanted Frenchmen instead, and intended to send a man to Canada to recruit French-speaking citizens.[130] Only news that his father had arranged to send some workmen from France prevented him from doing so. He did attract a small group of "Frenchmen and Canadians" which he valued highly,[131] but over the years Irénée adjusted to a largely Irish crew. In 1807, when an explosion blew up the dryhouse, Irénée was grateful that his employees "have not tried to leave me, as I expected of them, but have to the contrary redoubled [their] zeal."[132]

Eventually he acquired a trained group of men through accumulated experience, and came to regard them as excellent workers. Several years after beginning production he stated that:

> We have an excellent *Set* of workers, all peaceful and good workers. I have trained them myself, and commend myself in having been able to succeed with them, as a good worker is a precious thing and very rare in all industries in every country, especially this one.[133]

[128] J. Thomas Scharf, *History of Delaware, 1609–1888,* 2 vols. (Philadelphia: L. H. Everts, 1888), 2: 885–86.

[129] These words were written, but crossed out, in a draft of a letter to his father: E. I. du Pont, 28 April 1803, to P. S. du Pont, L3. He told his brother that he would have much preferred French and Germans: E. I. du Pont, 26 September 1802, to Victor du Pont, W4.

[130] E. I. du Pont, 25 March 1803, to P. S. du Pont, L3 and W4.

[131] Ibid.

[132] E. I. du Pont, 30 August 1807, to Victor du Pont, W4.

[133] E. I. du Pont, 22 January 1809, to P. S. du Pont, L3. The word "Set" he wrote in English.

Other powder makers had the same opinion. In 1808 a new company was formed in Richmond, Virginia, for the manufacture of black powder. The proprietors put an advertisement in a Wilmington paper for a master powderman, and sent an agent to negotiate with one of du Pont's men who answered it. Through him they obtained some minor equipment and information about powder making. Irénée discovered what was going on and took legal action which forestalled the attempted "labor piracy." Three years later another black powder entrepreneur, from Washington, D.C., made a clandestine attempt to acquire a du Pont workman. This time one of his loyal employees informed Irénée what was happening, and the resulting publicity foiled the plans.[134] Obviously the Brandywine workforce was known for its high quality within just a few years of its commencement of powder making.

Related to Irénée's original displeasure with his workers was his attempt to bring experienced powder makers from France. The idea of such a transplantation probably occurred to him while he was there in 1801. He talked with at least two men about emigrating, one a cooper (barrel manufacture was an important part of a powder works[135]) and one a skilled powderman at the Isle St. Jean and Essonne powder works.[136] Their willingness encouraged him to talk with officials of the *Administration des Poudres* who told him that they would send him such workmen as they could spare. As soon as Irénée returned to the United States he reminded them of their promises, especially requesting that they provide him with a master powderman but also some "ordinary workmen."[137] Irénée asked his father, who had returned to France, to act as his agent in this business, providing him with instructions as to the kind of contract he wanted with the Frenchmen.

> . . . it is indispensable that you have from each of them and also

[134] George H. Gibson, "Labor Piracy on the Brandywine," *Labor History* 8 (Spring 1967): 176–81.

[135] E. I. du Pont, 25 January 1802, to P.-M.-C. Robin, L3; cf. Wilkinson, "The Founding," 52.

[136] A. Bottée, 15 April 1802, to E. I. du Pont, L3; P.-M.-C. Robin, 9 May 1802, to E. I. du Pont, L3; E. I. du Pont, 16 July 1802, to A. Bottée, W4.

[137] E. I. du Pont, 25 January 1802, to Riffault, L3; E. I. du Pont, 25 January 1802, to P.-M.-C. Robin, L3; P.-M.-C. Robin, 24 June 1801, to E. I. du Pont, L3; E. I. du Pont, 16 July 1802, to A. Bottée, W4; E. I. du Pont, n.d. [May 1801], to Bottée, L2.

> the master workers an agreement of six to seven years for which there will be accorded to each of the latter 1,800 or 2,000 or 2,200 livres according to what the Messrs. [i.e., the officials of the *Administration des Poudres*] agree with them, and to the common workers about 800 *livres,* on the condition that the advances made to them for their travel will be returned little by little from their salary, and that they agree in the most formal manner not to leave without a month's notice, if they want to return [to France]. And that it is never to be possible to work in another powder factory in America under penalty of paying to this party 2,000 dollars indemnity for the master worker and 600 each for the common workers.[138]

But it was some time before any worker got near to signing such a contract. First, the two men with whom Irénée originally talked backed out of the agreement, and then another worker who had seemed ready to go changed his mind.[139] Finally, Bottée ordered four workmen to leave for America, and countermanded their orders only when Charles François Parent of Mayence agreed to go.[140] Parent was a master powderman who came highly recommended as exactly what Irénée needed.[141] Even P. S. du Pont was highly impressed, and as soon as Parent came to Paris he had him sign a contract with E. I. du Pont de Nemours and Co. in the presence of the American consul. It included exactly what Irénée had asked his father to specify, and gave Parent a salary of $500 a year for nine years.[142]

In April 1803, Charles François Parent, with his two sons and a governess, sailed from Le Havre to the United States. He arrived at New York and was met by Victor du Pont, who immediately sensed that something was wrong. He told his brother that Parent had acted "too much the gentleman" and might not be suited for American life.[143] After Parent reached

[138] E. I. du Pont, 15 August 1802, to P. S. du Pont, L3.

[139] P.-M.-C. Robin, 9 May 1802, to E. I. du Pont, L3; P. S. du Pont, 3 February 1803, to E. I. du Pont, L1; P. S. du Pont, 18 February 1803, to E. I. du Pont, L1.

[140] P. S. du Pont, 2 January 1803, to E. I. du Pont, L1; A. Bottée, 14 March 1803, to E. I. du Pont, box 2, L5; C. F. Parent, 6 February 1803, to A. Bottée, box 2, L5.

[141] Bussy and d'Allemagne, 18 March 1803, to E. I. du Pont, box 2, L5.

[142] P. S. du Pont, 18 March 1803, to E. I. du Pont, L1; folder on C. F. Parent, box 3, L1; file 21, Acc. 146, HML.

[143] P. S. du Pont, 13 April 1803, to E. I. du Pont, L1; E. I. du Pont, 21 May 1803, to Victor du Pont, W4.

Irénée's half-constructed establishment on the Brandywine other problems arose. Parent objected to the isolated and uncivilized locality in which he would have to work, wishing rather to be in Philadelphia or New York, which he thought were more beautiful than any city in France.[144] Nor did he like the condition of the powder works, in which he could not imagine himself working.[145] But most of all he objected to his salary which be believed to be far below that which should have been paid to a man of his position in America.[146]

This final complaint of Parent's is the only one which seems justified. Irénée had already commented on the high wages of American workers and believed that imported Frenchmen would be content with less.[147] But he was wrong; perhaps he did not count on the infectiousness of the American "get-rich" spirit. In any case, Irénée refused to increase Parent's salary, and the immigrant became even more dissatisfied. Parent was never any help, seldom leaving his house to look at the work in progress, and unwilling to give Irénée advice when asked.[148] Yet Irénée could scarcely order him to work while the equipment was not yet ready for use; Parent had been brought over as a workman, not a construction engineer.

Irénée was patient with Parent, although he had a growing dislike for the man. Finally, in February 1804, with production imminent, he decided to bring matters to a head. As he told Victor:

> I therefore wrote to him in order to know his intentions, persuaded that I could not proceed with him. He responded that the country did not suit him, and it was impossible to live with his salary. He asked nothing except that I send him back to France. I accepted immediately and it was then that he wrote to me to ask for passage for four persons on the *Amelia*, but do you believe that the same day, seeing his plan foiled by my acceptance, he told me that he had reconsidered and because

[144] C. F. Parent, 4 May 1804 [?], to Auguste Bottée (copy in Peter Bauduy's hand), L5: P. S. du Pont, 7 October 1805 [12 October 1804?], to E. I. du Pont, L1.

[145] E. I. du Pont, 8 August 1803, to P. S. du Pont, L3; E. I. du Pont, 12 June 1803, to P. S. du Pont, L3.

[146] E. I. du Pont, 17 March 1804, to Victor du Pont, W4; C. F. Parent, 4 May 1804 [?], to Auguste Bottée (copy in Peter Bauduy's hand), L5.

[147] E. I. du Pont, 12 March 1802, to J. A. A. Bidermann, W4; E. I. du Pont, 25 January 1802, to P.-M.-C. Robin, L3.

[148] E. I. du Pont, 17 March 1804, to Victor du Pont, W4.

> he was no longer useful to me he was going to establish himself in Philadelphia in order to use his skills to set up a distillery. I replied to him that since it was not the nation which had hired him, he must in that case hold to his contract with me, that it was in my works that his skills must be employed, and I asked him to come immediately to work with me. He responded in an evasive manner and when I renewed my demand he flatly refused and left for Philadelphia. . . . I had him taken to the prison at Newcastle [sic], prosecuting him for indemnities for the expenses which he has occasioned me, and I will have him held there until I have him agree to return to France or sign for me an agreement by which he must certainly be bound so that it is impossible for him to work in any manner to make powder in this country.[149]

Parent spent several months in jail before he and Irénée reached a mutually satisfactory agreement by the terms of which Parent was to be allowed to go to New Orleans to set up his own powder works.[150] Irénée gave him about a thousand dollars worth of equipment and supplies, and set up a partnership between Parent and Charles Dalmas, Irénée's brother-in-law. In addition he sent four hundred dollars worth of his powder for Parent to deliver to commission merchants in New Orleans. Despite fears that there might be more problems with this man who had already caused Irénée so much difficulty, Parent was able to establish himself successfully in the new territory.[151] Unfortunately he drowned accidently late in 1811.[152]

The problems caused by the Parent affair, while they may have had a significant effect on Irénée's ability to concentrate on his work, were diversionary rather than restrictive. A more pressing problem was the rapid diminution of the original capital of his enterprise, which forced Irénée to depend upon bank credit and delayed payments to allow completion of the works.[153] Just at the time he was ready to begin production

[149] Ibid.

[150] E. I. du Pont, 6 July 1804, to Victor du Pont, W4.

[151] Ibid.; E. I. du Pont, 12 October 1804, to P. S. du Pont, L3; E. I. du Pont, 11 November 1804, to Victor du Pont, W4; E. I. du Pont de Nemours and Co. (hereafter EIdPdN&Co.), 1 February 1811, to E. M. Ledet (includes account of and letter to "Mr. Chas. Parent & Co."), L5.

[152] *Veuve* Parent, 18 August 1812, to E. I. du Pont, box 9, L5.

[153] E. I. du Pont, 17 May 1803, to Victor du Pont, W4; E. I. du Pont, 22 December 1803, to Victor du Pont, W4.

Irénée wrote to his brother that he owed $4,500 to the bank and had only $350 to $400 in his account.[154] Less than two months later he was in desperate straits because he did not have the money to pay his workers.[155] Even after he had large sales of his first few months' powder, he was financially embarrassed.[156] The bank debt was long-standing, and in later years Irénée referred to it in a casual manner,[157] but during the two years of construction his finances were a serious problem and must have resulted in some slowing of work.

Illness was also a nagging problem of some importance. In the first few months of construction Irénée's workforce was constantly afflicted with the ague. He told his brother that "the ague is a common illness here so that I always have two or three workers here who are trembling."[158] As related earlier, his master mason got jaundice and his appearance frightened away the journeyman masons, which retarded construction. Irénée had to scour the countryside to get replacements, but by the time he got them freezing temperatures made it unlikely that they could continue work.[159]

Irénée was also ill, although he did not describe it as the ague. From the start of the work in the late summer of 1802 until the late spring of 1803, he complained of a disorder which he felt was in some way connected with mental strain.[160] He told his brother as he was ready to recommence construction in March of 1803:

> I need to stage a very great activity in my works, for I still have a terrible lot of things to do, and it is absolutely necessary for our establishment to march on as soon as possible. I am not nearly so far along as I had hoped to be this spring. The duration

[154] E. I. du Pont, 17 March 1804, to Victor du Pont, W4.

[155] E. I. du Pont, 30 April 1804, to Victor du Pont, W4; E. I. du Pont, 2 May 1804, to Victor du Pont, W4.

[156] E. I. du Pont, 12 October 1804, to P. S. du Pont, L3.

[157] E. I. du Pont, 2 May 1807, to P. S. du Pont, L3; E. I. du Pont, 8 February 1808, to P. S. du Pont, L3.

[158] E. I. du Pont, 18 October 1802, to Victor du Pont, W4.

[159] E. I. du Pont, 5 November 1802, to Victor du Pont, W4.

[160] E. I. du Pont, 26 September 1802, to Victor du Pont, W4; E. I. du Pont, 15 January 1803, to Victor du Pont, W4; E. I. du Pont, 7 February 1803, to Victor du Pont, W4; E. I. du Pont, 29 February 1803, to Victor du Pont, W4; E. I. du Pont, 28 April 1803, to P. S. du Pont, L3; E. I. du Pont, 12 June 1803, to P. S. du Pont, L3.

of the winter and my illness have kept me from doing as many things as I believed I could do.[161]

Each of the problems mentioned above—untrained laborers, language difficulties, the Parent affair, and illnesses—were factors in the long delay between the beginning of the construction of the du Pont powder works in July 1802, and the commencement of production in April 1804. A sufficient cause for such a delay, however, may have been Irénée's inexperience in the technique of building, as opposed to operating, a powder works.[162] While an *élève* in the *Régie* he had written an essay on the construction of powder mills, but it may not have been the result of observations of construction-in-progress.[163] (Although the inevitable explosions of powder mills did result in almost continual reconstruction.[164]) Irénée's writings after he had decided to build a powder works in the United States, and before beginning it, do not show an interest in construction problems. Soon after the start of building, however, he realized his deficiency, "I must admit that as well as never having had an idea of the rules of architecture, I have more difficulty than someone else, and especially more anxiety, when I realize that something is going badly due to my mistake."[165] This seems to have been the one skill contributory to success which Irénée did not bring with him from France, and it nearly caused his failure. He underestimated both the time and expense necessary for construction and was near emotional and financial exhaustion when production finally began. But early in April of 1804 Irénée reported to Victor that all his mill machinery was finally operating and that he expected to have some powder finished by the end of the week.[166] The company records show that the first sale—of twenty-five kegs—was made the next month.[167]

By July Irénée was more than relieved—he was exuber-

[161] E. I. du Pont, 29 March 1803, to Victor du Pont, W4.

[162] Cf. Wilkinson, "The Founding," 25.

[163] "Mémoire sur la construction des moulins à poudre," Acc. 37 and Acc. 519, HML.

[164] Grimaux, ed., *Oeuvres de Lavoisier,* 5: 741–45, describes an accident at Essonne in 1788, which may have been typical of those with which Irénée was familiar while an *élève.*

[165] E. I. du Pont, 18 October 1802, to Victor du Pont, W4.

[166] E. I. du Pont, 8 April 1804, to Victor du Pont, W4.

[167] "Accounts: powder sales," box 17, L5.

ant—as he stated that "until now we are selling as quickly as we produce and it is not even enough for the market."[168] A few months later he boasted to his father about his success.

> My business is going very well and presents a more favorable result than we ever could have hoped. My powder is not only infinitely above that of other manufacturers of the country, but it is even several degrees above the imported English powder. Moreover it sells as rapidly as it is possible to make it. . . .[169]

Soon he had better news. Thomas Jefferson wrote him a personal note assuring him that he would receive orders from government agencies which purchased powder.[170]

Irénée's claim that his powder was better than English powder was an assertion of superiority which could be understood by anyone familiar with the American powder market. The imported powder had established a high standard against which merchants immediately measured his.[171] The first reaction as it was sent out to merchants in 1804 was that du Pont powder was clearly better than average,[172] but also that it lacked the quality of "quickness"—rapid detonation when fired.[173] There were also complaints about lack of glazing, size of the grain, tendency to foul firearms, and the letter designations of the grading system which Irénée used.[174] Of course, Irénée could

[168] Sophie du Pont, 28 July 1804, to Victor du Pont, with postscript by E. I. du Pont, Series C, W4.

[169] E. I. du Pont, 12 October 1804, to P. S. du Pont, L3.

[170] Thomas Jefferson, 23 November 1804, to E. I. du Pont, file 70, Acc. 501, HML.

[171] "The imported . . . is of a remarkable [sic] good quality . . . ," George West, Jr., 1 October 1804, to Peter Bauduy, L5; ". . . English powder, tho [sic] highly glazed is commonly very quick, and all powder to meet with approbation at this market must be so." Mitchel and Sheppard, 25 March 1805, to EIdPdN&Co., L5; "So far as I have had Opp[ortunit]y of comparing it with British Powder, [the du Pont powder] has been approved . . . ," John Chew, 3 December 1805, to EIdPdN&Co., L5; "La poudre nous paraît belle, & les qualités bien distinguées . . . elle est supérieure aux poudres Européens." ["Your powder appears good to us, and its qualities very outstanding . . . it is superior to European powders."], Delaire and Canut, 18 March 1806, to EIdPdN&Co., L5.

[172] Mitchel and Sheppard, 13 June 1804, to Peter Bauduy, L5; Isaac Freeman, 19 July 1804, to Peter Bauduy, L5; Joseph Barber, 14 August 1804, to Peter Bauduy, L5; George West, Jr., 1 October 1804, to Peter Bauduy, L5.

[173] Mitchel and Sheppard, 27 December 1804, to Peter Bauduy, L5; John Mason, 5 February 1805, to Peter Bauduy, L5; Mitchel and Sheppard, 25 March 1805, to EIdPdN&Co., L5.

[174] (Glazing) ibid.; (size of grain) Isaac Freeman, 19 July 1804, to Peter Bauduy, L5; John Chew, 13 November 1804, to Peter Bauduy, L5; (fouling) John Chew, 1

not please everyone, but within two years merchant response to his product became almost uniformly favorable.[175]

In 1806 one merchant in Providence reported that:

> I am nearly out of Powder of your make & have established . . . [its] Credit in this State beyond what I could have expected considering Prejudice that prevailed, in this & the ajacent [sic] Towns against the American Powder owing to the grate [sic] Impositions [?] from Philadelphia & Connecticutt. . . .[176]

The United States military was less enthusiastic. Although Irénée was extremely pleased with Jefferson's intent to give him the War and Navy departments' business, he found it very difficult to meet their expectations for quality. He was given some old powder from the Philadelphia Arsenal to remanufacture in 1805, and after remaking it he tested it himself and found that it was of acceptable proof.[177] When received and tested at the Arsenal, however, it was found deficient. Irénée protested vehemently against the accuracy of the Arsenal's tests,[178] and began a long campaign to educate the military in what he believed to be the correct methods of proving powder.[179]

A major problem was that the military evidently did not establish consistent standards for powder testing. In June 1805, the superintendent of military stores in Philadelphia informed Irénée that to be acceptable one ounce of his powder must throw a twenty-four-pound ball sixty-five yards.[180] Yet two months later the secretary of war stated that the mortar *éprouvette* at Philadelphia was so imperfect as to render its tests useless.[181] In May 1806, when the state of Pennsylvania want-

November 1805, to EIdPdN&Co., L5; (grading) EIdPdN&Co., 18 December 1806, to Anthy. Charles Cazenove, Lb 1, Acc. 500; EIdPdN&Co., 27 November 1806, to Aubin La Forest, Lb 1, Acc. 500.

[175] Cf. Aubin La Forest, 10 October 1805, to EIdPdN&Co., L5; W. Addams & Co., 20 January 1806, to EIdPdN&Co., L5; Watkinsons and Co., 21 November 1806, to EIdPdN&Co., L5; William Rogers, 18 June 1807, to EIdPdN&Co., L5.

[176] John Whipple, 11 August 1806, to EIdPdN&Co., L5.

[177] E. I. du Pont, 1 July 1805, to General Dearborn, file 70, Acc. 501.

[178] E. I. du Pont, 23 August 1805, to General Dearborn, file 70, Acc. 501.

[179] E.g., H. Dearborn, 25 February 1806, to EIdPdN&Co., L5; E. I. du Pont, 21 May 1807, to General Dearborn, file 70, Acc. 501; E. I. du Pont, 26 January 1822, to Col. Wm. H. Tanner, L3.

[180] Callender Irvine, 11 June 1805, to E. I. du Pont, L5.

[181] H. Dearborn, 26 August 1805, to E. I. du Pont, file 70, Acc. 501.

ed to establish powder standards for its "Inspector of Gun-Powder" it was told that the government standard was sixty yards with one ounce of powder and a twenty-four-pound ball.[182] Six years later the government arsenal in Springfield, Massachusetts, required an ounce of powder to throw a twenty-three-pound ball eighty yards.[183]

Irénée attempted to counteract this official impreciseness in two ways. One was by conducting his own tests at Eleutherian Mills to satisfy himself of the quality of his product.[184] The other was by having outside tests made and the results sent to the War Department. The earliest of these was in July 1804, in Philadelphia, where Archibald McCall found that the du Pont powder performed better in a mortar test than English powder.[185] In a later one on Governor's Island in New York harbor the du Pont powder ranked above the English and equivalent to the Dutch.[186]

The apparent reason for the commercial—if not military—success which the du Pont powder enjoyed was the strict attention which Irénée paid to all phases of powder manufacture. He constantly sought to purchase the best saltpeter on the market, and tested what was offered him. He was especially careful to avoid saltpeter adulterated with sea salt, preferring to pay more for that with low levels of impurities. Irénée wanted saltpeter imported directly from India if possible, although he recognized the potential of saltpeter derived from cave deposits in Kentucky.[187]

Irénée also took a serious interest in the quality of the sulfur that he used. In August 1804, he wrote to his brother about its purchase, telling him to find

[182] *Pennsylvania Archives,* 9th ser., 10 vols. ([Harrisburg]: Department of Property and Supplies, 1931), 3: 2261–62. The testing device was to be at a 45-degree elevation.

[183] John Chaffee, 27 November 1812, to Callender Irvine, box 108, RG 92, National Archives, Washington, D.C. (microfilm of RG 92, reel 5, HML).

[184] Cf. E. I. du Pont, 1 July 1805, to General Dearborn, file 70, Acc. 501.

[185] E. I. du Pont, 6 July 1804, to Victor du Pont (W4-221), W4.

[186] Copy in R. du Planty, 23 May 1806, to EIdPdN&Co., L5; original in RG 107, National Archives, Washington, D.C. (microfilm of RG 107, reel 1, HML).

[187] E. I. du Pont, 1 April 1805, to Victor du Pont, W4; E. I. du Pont, 14 March 1805, to H[enry] Dearborn, L5; E. I. du Pont, n.d. [September 1807], to Anthony Girard, L3; EIdPdN&Co., 19 April 1806, to Anthony Girard, Lb 1, Acc. 500; EIdPdN&Co., 1 May 1806, to Anthony Girard, Lb 1, Acc. 500; EIdPdN&Co., 19 February 1808, to Archibald McCall, Lb 2, Acc. 500, HML; EIdPdN&Co., 1 March 1808, to Anthy. Girard, Lb 2, Acc. 500.

> and purchase some Italian sulfur, the yellowest and prettiest possible. That which you bought for me last year was excellent. Bauduy bought some at Philadelphia which had probably been refined here, the color of which was slightly brown, which I have no doubt, diminished the strength of my powder.[188]

In spite of these careful instructions, Irénée found that much of the sulfur available for purchase was unacceptable. So he went into the sulfur preparation business to insure a high quality ingredient for his powder, as a company correspondent was informed in December 1805.

> Having found that a great deal of the sulfur flour imported to this country was badly prepared and contained a great quantity of sulfuric acid which is very injurious to that medicine [sic] we have undertook to prepare it and find prity [sic] good sale for it in Philad. & Newyork [sic]. Wishing to extend our sales as much as possible to compensate the expenses that our *apparatus* has cost to us we send you one H[undre]d. Please let us know how the thing answer [sic] and show samples of it to your apothecary. We invoice it as low as we can in order to encourage the sale of it.[189]

Quantities of purified sulfur above what was needed for powder were sent to several other merchants, and within a few months Irénée was selling substantial amounts of it.[190]

Since charcoal was the one ingredient which Irénée manufactured on his own premises, there is little in his letters or the company's papers concerning it. It may be noted, however, that he made sure that willow, which he thought superior to other woods, was used exclusively, and that in one instance he experimented with better methods of charcoal making.[191]

Assured that his ingredients were the best available, Irénée then made certain that they were processed carefully and skillfully. From letters which he wrote to his wife, Charles Dalmas, and his son Alfred, when Irénée was on business trips, we know the kind of strict attention to processing which Irénée

[188] E. I. du Pont, 22 August 1804, to Victor du Pont, W4.

[189] EIdPdN&Co., 11 December 1805, to Richard Bowden and Co., Lb 1, Acc. 500.

[190] EIdPdN&Co., 1 January 1806, to Vincent Bousat, Lb 1, Acc. 500; EIdPdN&Co., 27 December 1805, to John L. Sullivan, Lb 1, Acc. 500; EIdPdN&Co., 14 June 1806, to Aubin La Forest, Lb 1, Acc. 500.

[191] E.g., E. I. du Pont, 20 March 1806, to Sophie du Pont, W4; see below.

expected of others. While away in the fall of 1804 he wrote to Dalmas to:

> Do your best, and double your attention to prevent accidents, for I have had a bad dream tonight. Try not to stop the manufacture for need of charcoal, and have it roasted by David in the cylinders. If you have no one else for it he will do it very well. Have him roast it in the small furnace, so as not to interrupt the manufacture of the Eagle [the high quality sporting powder]. . . . The door of the small furnace is nearly useless. Fix it at once somehow so that it will be of use at least once more. . . . Take care of the drying of the powder. . . .[192]

From this and other letters of this type,[193] we may assume that a man who showed such concern when away paid great attention to his works when present.[194]

There was yet another reason for the product's success: Irénée du Pont made numerous attempts to improve the technique of powder manufacture in his own works. One way was through continued contact with France. During construction of Eleutherian Mills he imported from France a large boiler (for saltpeter refining), as well as an assortment of millstones for the pulverizing mill.[195] He also requested his father to send him more materials for the graining mill from France, and for some time Pierre struggled to make some parchment sieves on patterns provided by Auguste Bottée.[196] Pierre also sought

[192] E. I. du Pont, 24 October 1804, to Charles Dalmas, L3.

[193] E. I. du Pont, 10 December 1805, to Sophie du Pont, W4; E. I. du Pont, 27 October 1806, to Sophie du Pont, W4; E. I. du Pont, 16 July 1807, to Sophie du Pont, W4; E. I. du Pont, 10 May [1819], to Sophie du Pont, L3; E. I. du Pont, 5 July [1823], to Alfred du Pont, L3; E. I. du Pont, 8 July [1823], to Sophie du Pont, L3.

[194] See the comment by Eugene S. Ferguson in *L'Acquisition des techniques par les pays non-initiateurs* (Paris: Centre Nationale de la Recherche Scientifique, 1973), 450–52.

[195] (Boiler) P. S. du Pont, 3 June 1802, to E. I. du Pont, L1; P. S. du Pont, 12 July 1802, to E. I. du Pont, L1; E. I. du Pont, 30 July 1802, to [?], L3; E. I. du Pont, 30 July 1802, to Victor du Pont, W4; (millstones) E. I. du Pont, 7 February 1803, to P. S. du Pont, L3; E. I. du Pont, 25 March 1803, to P. S. du Pont, L3; P. S. du Pont, [18 February 1803], to E. I. du Pont, L1; P. S. du Pont, 13 April 1803, to E. I. du Pont, L1.

[196] P. S. du Pont, 7 October 1804 [12 October 1804?], to E. I. du Pont, L1; P. S. du Pont, 24 December 1804, to E. I. du Pont, L1; P. S. du Pont, [30 May 1805], to E. I. du Pont, L1; P. S. du Pont, 10 September 1805, to E. I. du Pont, L1. For the importance of parchment sieves, see Guttmann, *The Manufacture of Explosives,* 211.

horsehair canvas for sifting and arranged to have some sent.[197] Irénée made some inquiries of his own concerning French sources for horsehair canvas, some of which was intended for the sun-drying of powder.[198]

A major concern in Irénée's letters to Bottée was the acquisition of plans for a new graining machine which Bottée had recently invented. Irénée understood that the machine was a labor-saving device, and felt that he needed it in the American situation of high labor costs. When Bottée did not immediately respond to his request for the plans, he called on his father's assistance, and eventually Bottée sent the plans.[199] They may have been the basis for the graining machine which Irénée patented in 1804.[200]

There was also some further personal information sent from France to the Brandywine. During construction of the mills Pierre was in close contact with Bottée and others in the powder administration and frequently passed on the results of his conversations with them to Irénée.[201] That this indirect method of transfer was significant is questionable, since Pierre had an inveterate interest in almost everything and a desire to appear to be an expert in whatever interested him.

Printed material was probably the most frequent French contact. Irénée was anxious to have the latest writings of the *Administration des Poudres,* and in 1802 and 1803 requested a copy of its *Instruction* to regional officials.[202] His father sent him books from France on powdermaking at least once,[203] and his friend John Vaughan at the American Philosophical Society in Philadelphia (of which du Pont was a member)

[197] P. S. du Pont, 22 November 1803, to E. I. du Pont, L1; P. S. du Pont, 7 October 1804 [12 October 1804?], to E. I. du Pont, L1; P. S. du Pont, [30 May 1805], to E. I. du Pont, L1.

[198] E. I. du Pont, 1 August 1804, to Pouchet-Belmare [?], L3. Some horsehair canvas had been ordered during Irénée's 1801 trip to France—see "Goods ordered in France," box 7, L3, and E. I. du Pont, 1 May 1805, to Victor du Pont, W4.

[199] E. I. du Pont, 26 January 1802, to Bottée, L3; E. I. du Pont, 16 July 1802, to [Bottée], W4; E. I. du Pont, 25 March 1803, to P. S. du Pont, L3; Bottée, [15 April 1802], to E. I. du Pont, L3; Bottée, 19 January 1803, to E. I. du Pont, L5.

[200] See below.

[201] E.g., P. S. du Pont, [3 November 1802], to E. I. du Pont, L1.

[202] E. I. du Pont, 26 January 1802, to Bottée, L3; E. I. du Pont, 16 July 1802, to Bottée, W4; E. I. du Pont, 25 March 1803, to P. S. du Pont, L3 & W4; E. I. du Pont, 28 April 1803, to P. S. du Pont, L3.

[203] P. S. du Pont, 21 June 1803, to E. I. du Pont, L1.

wrote to him several times concerning European innovations in powdermaking of which he had heard.[204] In 1812 Irénée learned that Bottée and Riffault had finally published their book *Traité de l'art de fabriquer la poudre à canon* (Treatise on the Art of Making Gunpowder), and insisted that his father procure him copies of it. He was concerned that other American manufacturers would take advantage of its contents before he could.[205] In the late 1820s he had an experienced French workman write an essay for him on the distillation process for making charcoal,[206] and when his son-in-law and business associate, Antoine Bidermann, traveled to France in 1827 and 1828, he was instructed to learn as much as he could of new French processes.[207]

Irénée seldom, however, credited printed material or information in letters from France with improving his manufacture. Rather he frequently asserted that his improvements were the result of his constant experimentation.[208] Typical was his comment in a letter to Commodore David Porter in 1820 that "I think I had improved the manufacture of gunpowder here, more than it had heretofore been done in Europe."[209] There is little in existing records to bear out this boast, although there are some references which raise questions such as: what was the "discovery" which Irénée mentioned to his father in 1804 which would have radically reduced the time necessary for incorporation in the stamping mill?[210] And, did Irénée use steam heat in his new dry house built in 1807?[211]

[204] John Vaughn, 30 March 1804, to E. I. du Pont, L3; John Vaughn, 29 December 1803, to E. I. du Pont, L5; John Vaughn, 12 January 1804, to E. I. du Pont, L5; John Vaughn, n.d. [1811?], to E. I. du Pont, L3.

[205] E. I. du Pont, n.d. [1812], to P. S. du Pont, L3; Norman B. Wilkinson, "Brandywine Borrowings from European Technology," *Technology and Culture* 4 (Spring 1963): 9.

[206] EIdPdN&Co., 25 December 1827, to W. Kemble, Acc. 500.

[207] E. I. du Pont, 12 February 1828, to Antoine Bidermann, and E. I. du Pont, 29 April 1828, to Antoine Bidermann, file 132, Acc. 146, HML. Also see the letters of Bidermann to du Pont, 1827–28, box 6, W4.

[208] Note Wilkinson, "The Founding," 64, 67.

[209] EIdPdN&Co., 20 October 1820, to Commodore David Porter, letterbook for 1820–21, Acc. 500.

[210] E. I. du Pont, 12 October 1804, to P. S. du Pont, L3.

[211] E. I. du Pont, 8 February 1808, to P. S. du Pont, L3; James Means, 17 January 1808, to E. I. du Pont, L3; John Vaughn, 29 December 1803, to E. I. du Pont, L5; John Vaughn, 12 January 1804, L5.

Occasionally there is a record of something more substantial to indicate on what his claim was based. In 1807 Pierre du Pont learned from Count Rumford, the well-known expatriate American scientist, about the distillation method of making charcoal which had been developed a decade earlier in England, and wrote a short description of the process for his son.[212] A few months later Irénée related to his father his trials of the process.

> I have attempted twenty different ways of doing it and equally without success. The charcoal which I obtained was rather good, but, however, inferior to that made by the ordinary process used in France. I have regretted not having been able to succeed in it, since that way would be much more economical, but I have always found the charcoal heavier and denser, and I do not believe it possible, therefore, to remove entirely the oily part of the wood in a closed vessel. If you force it into the fire it will only become harder, and in consequence, less usable in the manufacture. I would be very glad to know of the trials that have been made on this subject by the officials of the powder administration, and what they think of it.[213]

Irénée also experimented with machinery. Shortly after receiving plans of Bottée's graining machine he reported to his father some of his experiments with reducing manual labor in the graining process.

> I have succeeded in inventing a machine which does as much work by itself as six men do with the graining machine with a handle; it does three-fourths of the grain instead of half as the ordinary types; it works in a box, much diminishing losses and dangers by avoiding the quantity of dust which now spreads in all the buildings as a result of procedure used now.
>
> This machine, of which I have made a test, succeeds completely, and I have asked for it a patent of fourteen years, which the laws of this country allow me to obtain. I have now made four of these machines which run by water, doing all the graining

[212] P. S. du Pont, 24 July 1807, to E. I. du Pont, L1.

[213] E. I. du Pont, 8 February 1808, to P. S. du Pont, L3. Cf. Bottée and Riffault, *Traité de l'art,* 1: 140: "The results of these tests on the carbonization process of the English, ought to prove to the *administration* that they are not preferable to those used in France, and to confirm it especially in the idea which it had of the ridiculous exaggeration of the pretended advantages of the modes of carbonization used in England."

of my works and saving me more than 2,000 dollars per year in labor and losses.[214]

The patent application was accepted, and entered on 23 November 1804.[215] It was the only patent which Irénée ever received.

Whereas there is less evidence that Irénée "improved" his manufacture through development of new processes and machines, there is ample evidence that he had a critical attitude toward both his raw materials and his product. Somewhere in his works, probably in the refinery, he had a laboratory which he partly equipped with French apparatus. On his trip to France in 1801 he had purchased "three mercury thermometers for chemistry with two rulers [scales] mounted in pearwood [and] eight areometers for nitre and salts with their cases."[216] The areometer, a device for determining the specific gravity of a solution of saltpeter and thereby the quantity of ocean salt mixed in, was indispensable in making high quality saltpeter.[217] It was a disaster when nearly all those he had purchased were broken *en route.* He wrote to a business associate in France to send a dozen more as soon as possible, and specified that they be those required by an *instruction* of the French *Administration.*[218] He also purchased some glassware in New York soon after returning from France.[219] All these items were mentioned by Bottée and Riffault in 1811 as equipment necessary for laboratory testing of saltpeter.[220] It was Irénée's laboratory that allowed him to make accurate assessments of saltpeter offered for sale.[221]

Quality control also required testing the end product. Irénée was always concerned to have powder-testing devices—*éprouvettes*—on hand. He purchased a small pendulum-type *éprouvette de Regnier* while in France in 1801 and ordered a dozen

[214] E. I. du Pont, 12 October 1804, to P. S. du Pont, L3; cf. E. I. du Pont, n.d. [ca. October 1804], to James Madison, L3.

[215] Henry L. Ellsworth, *A Digest of Patents Issued by the U.S. from 1790 to Jan. 1, 1839* (Washington, D.C.: n.p., 1840), 426.

[216] Receipt, 6 Floréal IX, L5.

[217] Bottée and Riffault, *Traité de l'art,* 1: 25–26.

[218] E. I. du Pont, 7 February 1802, to Harmand, Acc. 23, HML.

[219] Receipt, 9 September 1801, "1801" folder, box 7, L3.

[220] Bottée and Riffault, *Traité de l'art,* 1: 396–98.

[221] See above.

more after his return.[222] This device took a small charge of powder and, when fired, drove a swinging arm through an arc marked off into units up to thirty.[223] Although the units may have had an absolute value, Irénée used them as a basis of comparison when he wrote to his brother in 1804 that his powder gave "15 and 16 degrees on the éprouvette and none of the English powders which we have tested until now have given more than 13 to 14. You have, therefore, the guarantee of equality to the imported powder."[224] He felt that *éprouvettes de Regnier* were such accurate and useful instruments that he gave some to merchants who handled powder, and two to the War Department.[225]

It was with the War Department that Irénée had some difficulty in establishing standards for powder testing. He had obtained at least one mortar *éprouvette* from the *Administration des Poudres,* but the War Department had no such standard device.[226] Since the government was a major purchaser of powder, Irénée several times carefully explained to War Department officials the conditions under which powder could be accurately tested. It was basically a matter of keeping all the samples dry and then using an *éprouvette* appropriate for the grade of powder being tested.[227] The fine powder used for carbines had to be tested with a pistol *éprouvette,* whereas the coarse powder used for heavy ordnance had to be tested with a mortar *éprouvette.*[228]

The testing of powder to control its quality seems to have been an integral part of operations at Eleutherian Mills. Irénée used the tests, at least in the early years, as a measure of the

[222] "Dépense générale," box 15, L5; expense lists, February–May 1801, "1801" folder, box 7, L5.

[223] Bottée and Riffault, *Traité de l'art,* 2: 514–515, plate 37, fig. 2.

[224] E. I. du Pont, 23 July 1804, to Victor du Pont, W4.

[225] E. I. du Pont, 25 March 1803, to P. S. du Pont, L3 & W4; E. I. du Pont, 1 September 1803, to Victor du Pont, W4.

[226] E. I. du Pont, 25 March 1803, to P. S. du Pont, L3 & W4; P. S. du Pont, 21 June 1803, to E. I. du Pont, L1; P. S. du Pont, 7 July 1805, to E. I. du Pont, L1; P. S. du Pont, 2 September 1805, L1; H. Dearborn, 26 August 1805, to E. I. du Pont, box 2, Acc. 501; Bottée and Riffault, *Traité de l'art,* 2: 509–12, pl. 36.

[227] E. I. du Pont, 23 August 1805, to Dearborn, file 70, box 1, Acc. 501; E. I. du Pont, 13 June 1805, to Dearborn, file 70, box 1, Acc. 501; [E. I. du Pont], 20 February 1806, to Dearborn, L5; R. du Planty, 23 May 1806, to EIdPdN&Co., L5.

[228] E. I. du Pont, 26 January 1822, to Col. Wm. H. Tanner, L3; E. I. du Pont, 26 October 1807, to Dearborn, L5.

potential competitiveness of his powder on the market.[229] He also used used it as a check on the tests made by the War Department of the powder he delivered to them.[230] Both kinds of tests indicate the interest and pride which Irénée had in his powder.

In assessing the transfer of French gunpowder technology to the Brandywine, the most apparent and striking fact is that so much depended upon one man, E. I. du Pont. He had the training and knowledge required to set up a powder works. He knew how to purchase equipment, select a site, direct construction (with some difficulty), and supervise production. The first and the last of these abilities were unique in the United States and were due to Irénée's training at the *Régie des Poudres.* No other American could have said, as he did, that "I have always regarded myself as an offspring of the powder administration."[231]

If no American had the skills to do what he did, none had the opportunity to import French plans and equipment, either. Although much machinery was built on the Brandywine and some equipment was designed by Irénée, the many items imported from France, particularly chemical and powder-testing apparatus, were important to his success. The quality control which Lavoisier had been instrumental in introducing into French powder manufacture required fine equipment which was normally unavailable in the United States.

Aside from his French technical training and sources of equipment, Irénée brought to the United States other elements important to his success: business training and acumen. The years which he spent in partnership with his father in the print shop provided valuable knowledge which allowed him to step into the role of entrepreneur rather easily. The exhaustive negotiations with Broom over the Brandywine millsite, the ability to continue construction in the face of numerous difficulties, and the rapid success of the enterprise once in production indicate that he was a capable businessman.

[229] E. I. du Pont, 6 July 1804, to Victor du Pont, W4; E. I. du Pont, 23 July 1804, to Victor du Pont, W4; EIdPdN&Co., 8 September 1806, to Mitchel & Sheppard, Lb 1, Acc. 500.

[230] E. I. du Pont, [7 August 1816], to Sophie du Pont, L3; E. I. du Pont, 23 August 1805, to [Henry Dearborn], file 70, Acc. 501.

[231] E. I. du Pont, 24 April 1807, to DeBussy, L3.

The story of Eleutherian Mills after the production of the first powder until Irénée's death thirty years later is one of cautious growth, but not prosperity. Profits were mainly swallowed up by Irénée's insistence on reinvesting as much as possible in new buildings, better equipment, and more land along the Brandywine. He wanted to make his establishment the best of its kind in America, and his powder the highest in reputation. Peter Bauduy, his American partner, disagreed with that approach, believing that expanding sales and profits should be the measures of success. In 1815 after a bitter struggle Irénée forced Bauduy out of the company and continued his policies of self-reliant growth.

What remained of the company's profits during Irénée's years was snapped up by the banks for interest on loans (which were particularly necessary after a disastrous explosion in 1818), and by the company's stockholders in France. In 1814 the stockholders even sent a representative to America to investigate the lack of cash profits, but his report of the company's enormous growth in net worth restored their confidence. At the time of his death in 1834 Irénée had made substantial progress in paying off his bank loans, and had repatriated many of the French shares, but still had many financial obligations. On the other hand, Eleutherian Mills was by far the United States' largest powder mill and, according to the policies of its founder, it produced black powder of the highest quality.[232]

The du Pont gunpowder works was a successful transfer of technology in its own right, but what place does it have in the historic process of the transfer of gunpowder technology to the United States? Lacking a good history of gunpowder manufacture in the United Sates, it is impossible to be definite, but it appears that Irénée's transfer was only one of many. The two powdermen from the *Régie des Poudres* who came to America during the Revolutionary War, the Batavian at the Decatur and Lane works in Philadelphia, and Charles François Parent all made contributions. Undoubtedly there were others, requiring further research to identify, as the American gunpowder industry before Irénée's arrival was extensive.

[232] William S. Dutton, *Du Pont: One Hundred and Forty Years* (New York: Charles Scribners' Sons, 1942), 41–65.

The impact of E. I. du Pont's transfer of French gunpowder technology was also muted by his attempt to prevent the hiring of his workmen by others. Although his restrictive policies may not have been always successful, the Brandywine enterprise appears remarkable for its low turnover in skilled personnel. The only significant defection in the early years, besides Parent, occurred in 1814 when Peter Bauduy established the Eden Park powder mill.[233] Warren Scoville's assertion that the transfer of technology by a single individual is unlikely to lead to rapid diffusion is particularly applicable to the case of E. I. du Pont.[234]

Du Pont's transfer of gunpowder technology to the United States was a success due to his technological training, his business skills, and the importation of some essential equipment from France. Although it was only one in a series of transfers of gunpowder technology to the United States, and one which did not lead to a rapid diffusion of skills, it did result in the establishment of a large gunpowder works producing a high quality product. Irénée's experience reflects both the potential and the limitations of the transfer of technology by an individual.

[233] Van Gelder and Schlatter, *History of the Explosives Industry,* 85, 87, 187.

[234] Warren C. Scoville, "Minority Migrations and the Diffusion of Technology," *Journal of Economic History* 11 (Fall 1951): 350.

IV: Moncure Robinson and the Origin of American Railroad Technology

Sometime during the winter of 1824–25 Moncure Robinson decided to go to Europe. It was an important decision not only for his own life and career, but also for American technology, since he was one of the first American engineers to acquire firsthand knowledge of the state of the engineering profession in Europe. Moreover, he was one of the first Americans to see the beginning of the modern railroad age in Britain, and he became a major transferor of railroad technology to the United States.

Robinson was one of a group of at least fifteen Americans who, having received fundamental training and experience in engineering, traveled to England from 1825 into the 1830s in order to examine and understand British railroad technology. Some of these men were apprentices to early American civil engineers, others had "on-the-job" training with important internal improvement projects, and some had been educated at the United States Military Academy (West Point).[1]

These American engineers did not go to Europe under common auspices. Robinson, John Edgar Thomson, and Edward Miller went because of their own fascination with tech-

Key to Footnote Abbreviations

AFP = Ambler Family Papers, Alderman Library, University of Virginia, Charlottesville, Virginia
HML = Hagley Museum and Library, Greenville, Wilmington, Delaware
HSP = Historical Society of Pennsylvania, Philadelphia, Pennsylvania
MRP = Moncure Robinson Papers, Swem Library, The College of William and Mary in Virginia, Williamsburg, Virginia
PaCC = RG 17, Records of Land Office, Pennsylvania Board of Canal Commissioners, William Penn Archives, Harrisburg, Pennsylvania
VaBPW = Records of the Virginia Board of Public Works, Virginia State Library, Richmond, Virginia
VHS = Virginia Historical Society, Richmond, Virginia

---

[1] A list of these fifteen engineers is compiled in Darwin H. Stapleton, "The Origin of American Railroad Technology, 1825–1840," *Railroad History* 139 (Autumn 1978): 65–77.

nical innovation. Others were commissioned by interested organizations: the infant Baltimore and Ohio Railroad Company sent four of its engineers to England in 1828–29 because there was simply no one in America yet prepared to build a railroad.

While abroad these American engineers acquired knowledge of railroad technology through meetings with British engineers and close observation of the newest railroads. William Strickland became a good friend of Jesse Hartley, a prominent railroad engineer, during his visit of 1825, and Horatio Allen found both Hartley and George Stephenson (the father of modern railroads) "perfectly willing to converse on whatever topic I wished to introduce." Edward Miller and Wirt Robinson inspected the Liverpool and Manchester Railroad during their visits, and Jonathan Knight, William McNeill, and George Whistler also saw the Stockton and Darlington line. Apparently there was little to keep American engineers from seeing and asking about those things which interested them.[2]

These engineers all became advocates of railroad construction on their return to the United States, and each became the chief engineer of one or more railroads in the early years of American railroading. Many of the lines built by the returned Americans were imitations of the most impressive British railroad, the Liverpool and Manchester. In following that modern railroad so closely the American engineers did what has often been done in technology transfers: they brought from Europe the latest railroad technology, which was relatively untried, rather than an older version which had clearly understood capabilities and limits.

British railroads had Elizabethan origins. Migrant German miners probably brought the idea of the railed way from central Europe, but it took a new form in seventeenth-century Britain: coal wagons with flanged wheels ran on raised rails to carry coal from the pit to river barges. By the latter eigh-

[2] Agnes A. Gilchrist, *William Strickland: Architect and Engineer, 1788–1854,* enlarged ed. (New York: Da Capo, 1969), 6, 12; Robert E. Carlson, "British Railroads and Engineers and the Beginnings of American Railroad Development," *Business History Review* 34 (Summer 1960): 140, 145, 147–48; Solomon W. Roberts, "Obituary Notice of Edward Miller," *Proceedings of the American Philosophical Society* 12 (1871–72): 582; 30 December 1831, W. D. Lewis Diary, Historical Society of Delaware, Wilmington; Eugene S. Ferguson, ed., *Early Engineering Reminiscences (1815–1840) of George Escol Sellers* (Washington, D.C.: Smithsonian Institution, 1965), 134.

teenth century railroads with horse-drawn cars were a common sight in mining districts, and sometimes were over ten miles long. On some of these lines there were even cast-iron edge rails mounted on stone or timber sleepers, a clear step toward the modern rail and track.[3]

What brought the railroad to its final form was the development of a reliable locomotive. Historians usually credit Richard Trevithick of Wales with assembling the first steam locomotive (originally a road vehicle) in 1801, and cite the performance of the *Rocket* at the Rainhill trials of the Liverpool and Manchester Railroad in 1829 as the event which showed the superiority of steam locomotion over animal power for traction.[4] When the locomotive was combined with iron rails, the essential elements of the modern railroad were in place.

That combination was most visible on two lines, the Stockton and Darlington, completed in 1825, and the Liverpool and Manchester, completed in 1830. George Stephenson was the constructing engineer of both railroads. His success in building those lines established them as models of railroad design, and they were initially regarded as imbued with a correctness of form amounting to natural law.[5]

Important elements of the Stephenson system included iron edge rails set on parallel rows of square stone sleepers, grades of no more than about one-half percent (since the adhesive power of locomotives was thought to be relatively low), inclined planes with stationary engines for steeper grades, and steam locomotives. An overriding value in every phase of construction was "permanence"—the notion that it was best practice to build for the ages, preferably by depending upon stone and iron to create rigid structures. The Liverpool and

[3] John Geise, "What is a Railway?" *Technology and Culture* 1 (Winter 1959): 68–76; M. J. T. Lewis, *Early Wooden Railways* (London: Routledge and K. Paul, 1970), 89–90, 105–109.

[4] E.g., Eugene S. Ferguson, "Steam Transportation," in *Technology in Western Civilization*, eds. Melvin Kranzberg and Carroll W. Pursell, Jr., 2 vols. (New York: Oxford University Press, 1967), 1: 293, 297–98.

[5] Robert E. Carlson, *The Liverpool and Manchester Railway Project* (Newton Abbott: David & Charles, 1969); C. von Oeynhausen and H. von Dechen, *Railways in England: 1826 and 1827*, eds. Charles E. Lee and K. R. Gilbert, trans. E. A. Forward (Cambridge: Newcomen Society, 1971); L. T. C. Rolt, *George and Robert Stephenson: The Railway Revolution* (London: Longmans, 1960).

Manchester rail and track were, for example, called "the permanent (or 'perfect') way."[6]

The success of railroads in Britain led to serious consideration of them in the United States. American visitors to Britain, especially those concerned with technical and commercial matters, made critical assessments of what they saw, and many were so impressed they spread news of them among their friends upon their return. One instance of this was reported in 1824 by Josiah White, a civil engineer who had promoted and constructed the Lehigh Navigation in northeastern Pennsylvania. He wrote excitedly to a friend interested in developing the anthracite coal regions: "I am informed by Jonah Thompson who is a principal Owner of the Phoenix nail works on French Creek [near Pottstown, Penna.], that he has seen in England an Iron Rail Road of three miles in length, to Carry Coal by an *Endless chain the whole way.*"[7]

That sort of news was behind a transatlantic visit later in the same year by a Baltimore merchant, possibly an agent of Alexander Brown and Sons, to collect information on railroads. Shortly thereafter the Pennsylvania Society for the Promotion of Internal Improvement was founded in Philadelphia, and early in 1825 it sent William Strickland to Britain as its agent. The reports which he sent back were full of information about railroads, and he urged their construction where traffic was heavy and speed important. Other news of railroads was published, and most literate Americans must have been aware of railroad developments in England by the later 1820s.[8]

Another element fueling interest in railroads in the same decade was, paradoxically, the success of the Erie Canal and the opening of other canals, such as the Lehigh, Schuylkill, and Delaware and Hudson. These new transportation routes

[6] Carlson, *The Liverpool and Manchester Railway Project,* 198.

[7] Josiah White, 1 January 1824, to Jacob Cist, Jacob Cist Correspondence, Academy of Natural Sciences of Philadelphia. On Cist's promotional efforts, see H. Benjamin Powell, *Philadelphia's First Fuel Crisis: Jacob Cist and the Developing Market for Pennsylvania Anthracite* (University Park and London: Pennsylvania State University Press, 1978).

[8] Carlson, "British Railroads and Engineers," 138–41; Carlson, *The Liverpool and Manchester Railway Project,* 15; Julius Rubin, *Canal or Railroad?* (Philadelphia: American Philosophical Society, 1961), 22–26.

had a dramatic effect on the development of hitherto remote regions, especially the anthracite coal district of northeastern Pennsylvania. It should not be surprising that through analogy to British experience Americans looked to railroads to be "feeders" for the canals. More intangibly, the success of the canals added to the growing entrepreneurial, promotional spirit that since about 1815 had directed large sums of private and government monies to the construction of "internal improvements."[9] News of the Stockton and Darlington, and Liverpool and Manchester railroads came to an American people receptive to the idea of transport innovation.

One other element was necessary, however, for railroads to be built in the United States: a group of people trained in the art of planning and directing the construction of large public works—in other words, civil engineers. Before about 1815 there were few men in America who could claim to be members of that profession. Two English engineers, William Weston and Benjamin Henry Latrobe, had crossed the ocean and had not only directed some canal construction but had also trained some assistants (see chapter 2). One American, Loammi Baldwin, Jr., had visited Europe to obtain much of his civil engineering knowledge.[10] But only after 1815 was the first significant generation of American civil engineers educated and trained.

The United States Military Academy at West Point was founded in 1802, but it was not a strong educational institution until Sylvanus Thayer became superintendent. His visit to Europe in 1815–16, just before assuming that position, led him to install at the academy the French mode of technical education with a prescriptive, heavily scientific course. French or French-trained instructors in mathematics and engineering were the core of the faculty. Many of the graduates of West Point became civil engineers, although not all had to leave

[9] Charles Hadfield, *British Canals: An Illustrated History,* 4th ed. (New York: Augustus M. Kelley, 1969), 110–14; George Rogers Taylor, *The Transportation Revolution, 1815–1860* (New York: Harper and Row, 1968), 18–20, 25–26, 48, 52.

[10] William Weston's trainees included Benjamin Wright, Robert Brooke, and Loammi Baldwin, Sr. Robert Shelton Kirby, "William Weston and His Contribution to Early American Engineering," *Transactions of the Newcomen Society* 16 (1935–36): 111–25; Daniel Hovey Calhoun, *The American Civil Engineer* (Cambridge, Mass.: M.I.T. Press, 1960), 34.

the Army to do so. Congress provided in an act of 1824 that Army engineers assigned to the Topographical Corps could be assigned to state or even private works at the discretion of a Board of Engineers for Internal Improvement. Until 1838 a large number of Army engineers saw such non-military service, and it is interesting that in the service of states and private companies many of them visited Europe or worked under American civil engineers who had.[11]

The other great institution for training civil engineers was the New York canal system (the Erie and Champlain canals) on which construction began in 1817, although other states' canal projects which started later had a similar impact. Benjamin Wright, the engineer most prominent in early years of the New York system, had received some training under William Weston, the British engineer; Canvass White, whose influence grew throughout the construction period, visited Europe in 1817 to study canals. Under these two there was a vast structure of subordinates of various ranks who were promoted to more responsible positions as they developed their skills. Many of the engineers trained in this fashion went on to other state or private internal improvement projects.[12]

In addition to these two institutions, many civil engineers were taught their profession through apprenticeship or "on-the-job" training of a less structured nature than on the Erie Canal. Benjamin Henry Latrobe trained four engineers in that manner.[13] Through apprenticeship and on-the-job training, West Point, and the New York canals, there was a substantial supply of civil engineers by the time railroads were seriously contemplated in the United States.[14]

American civil engineers were not, however, sufficiently trained to build railroads as a result of their American experience. In order to acquire knowledge of railroads they had to cross the Atlantic to see them in operation. Moncure Robinson was one of the first American civil engineers to see and

[11] Calhoun, *The American Civil Engineer,* 37–43; Forest G. Hill, *Roads, Rails, and Waterways: The Army Engineers and Early Transportation* (Norman: University of Oklahoma Press, 1957), 105.

[12] Calhoun, *The American Civil Engineer,* 24–30, 34–37.

[13] Ibid., 47–48; Darwin H. Stapleton, ed., *The Engineering Drawings of Benjamin Henry Latrobe* (New Haven: Yale University Press, 1980), 68–69.

[14] Calhoun, *The American Civil Engineer,* 50–53.

understand the technology of modern railroads in Britain, and as a result he went on to survey and build a number of early railroads in Virginia and Pennsylvania, including the Philadelphia and Reading Railroad—one of the finest examples of engineering skill in early American railroad history.

Moncure Robinson was born in 1802 to a merchant family of Richmond, Virginia. His parents were concerned about his education and sent him to a tutor where, among other subjects, he was taught French.[15] At age thirteen he entered the College of William and Mary, where he was a diligent student, although he left in 1818 without receiving a degree.[16] Sometime while in college or shortly thereafter he decided to begin a career in surveying or, perhaps, civil engineering, because in 1819 he volunteered to be an assistant, without pay, on a state expedition headed by Thomas Moore to explore the possibility of constructing a canal to connect the James and Kanawha rivers. This arduous service of about seven months earned him special notice in the expedition's official report to the Virginia Assembly.[17]

Robinson's work under Moore first brought him into contact with the nascent American engineering profession. Moore

[15] Receipts for tuition, 1810 and 1815, MRP; Moncure Robinson, 10 February 1826, to John Robinson, MRP; [Richard Boyse Osborne], *Sketch of the Professional Biography of Moncure Robinson, Civil Engineer* (Philadelphia: J. B. Lippincott Company, 1888), 2. (Osborne's biography was reprinted in the *William and Mary Quarterly*, 2nd ser., 1 [1921]: 237–60.) Richard Boyse Osborne was a subordinate to Moncure Robinson on the Philadelphia and Reading Railroad, and later pursued an engineering career in the Philadelphia area, where Robinson was a resident. This biography, written three years before Robinson's death, was probably the result of their personal acquaintance. There are, however, numerous dating errors in the text, and some other inconsistencies with other sources, so it has been used sparingly as a source for this chapter.

[16] John Robinson, 18 December 1815, to Moncure Robinson, MRP; Martha H. McGill, 10 November 1816, to John Robinson, MRP; Ferdinand S. Campbell, 26 February 1817, to John Robinson, MRP; J. Augustus Smith, 4 March 1818, to John Robinson, MRP.

[17] "Report of the Principal Engineer . . . ," *Fourth Annual Report of the Board of Public Works* (Richmond: Thomas Ritchie, 1820), 113–14; receipts for expenses of Moncure Robinson, 1819, (uncatalogued) Moncure Robinson box, VaBPW; Wayland Fuller Dunaway, *History of the James River and Kanawha Company* (New York: Columbia University, 1922), 64–65.

Thomas Moore (1760–1822) was a surveyor, inventor, and engineer. In 1806 he was appointed one of the original commissioners of the National Road, and from 1818 to 1822 he was the state engineer of Virginia. Calhoun, *American Civil Engineer*, 31, 36; Thomas B. Searight, *The Old Pike: A History of the National Road* (Uniontown, Penna.: published by the author, 1894), 28; Daniel J. Boorstin, *The Americans: The Democratic Experience* (New York: Vintage Books, 1974), 328–39.

Fig. 9. Moncure Robinson, ca. 1840. Courtesy of Hagley Museum and Library.

is a good example of the native American engineer of his era: he was trained by practical experience (largely in surveying) rather than by education or apprenticeship, and he valued the goals of promotion in planning and expediency in construction. The differences between Moore's approach and that of British engineers were made explicit in his public debate of 1811–12 with Benjamin Henry Latrobe over the improvement of the Potomac River at Georgetown. Merchants wanted to improve the town's harbor, which was virtually blocked by silt.

Moore advocated using wooden wingdams to concentrate the river's flow, consequently scouring out the silt and permitting ships to dock at Georgetown immediately. Latrobe cited European experience which showed that the silt would soon accumulate in a new site farther downstream, blocking the river once more and making the investment in the wingdams worthless. Latrobe's British training made him favor only those projects which could be soundly built of permanent materials, and which would provide long-term solutions.[18]

The Moore-Latrobe conflict exemplified the difference between American and British approaches to engineering. Robinson was to be part of the new generation of American engineers which uniquely combined the traditions. Robinson embarked on a career as a railroad engineer after he examined British railroads firsthand and understood their principles of design and construction, but his initial exposure to Moore's American principles served him equally well in helping to develop uniquely American railroads.

Robinson's life after serving in Moore's surveying expedition is not well illuminated by surviving records, but it may be that he used the experience gained on the expedition to work as a surveyor. In one letter from this period Robinson apparently referred to himself when he suggested that an "accurate Surveyor" should be appointed to an opening with the Virginia Board of Public Works. His skills were also put to use when, as a biographer later reported, Robinson made a second surveying trip to the western part of Virginia, but this time alone and "for the purpose of hunting up and locating his father's wild lands."[19]

In the summer of 1822 the young Virginian traveled to New York to examine the Erie Canal and talk with its engineers. He saw the full length of the Erie, made drawings, and reported to his father that: "I reached the Canal at the season when I had the best opportunity of seeing work in every state of their progress. I have perhaps acquired as much informa-

[18] Stapleton, ed., *The Engineering Drawings of Benjamin Henry Latrobe,* 42–44.

[19] Moncure Robinson, 22 January 1821, to John Robinson, MRP; Osborne, 6–7.

tion as I could, without being actually engaged in their superintendence."[20] Robinson's purpose in seeing the Erie was to become better qualified for a position with his own state's James River (canal) Company. That company had been in operation for some years without accomplishing very much, but after the state government assumed control of the company in 1820 it looked as if some construction would take place.[21] In 1823 work was begun on reconstructing a section of the canal from the falls of the James River at Richmond to a location upriver known as Maiden's Adventure, and Moncure Robinson was appointed the engineer.[22]

Robinson directed the construction of a number of structures common to canals, such as locks, culverts, bridges, and the canal trough itself. He was engaged on this work for two years until it was virtually completed, and when he left he was the subject of an adulatory report by the state commissioner to whom he was responsible.[23] But he was not satisfied with his situation in Virginia, partly because it did not look as if the state was likely to fund any further internal improvements,[24] and partly because he was enthralled by the possibility of going to Europe.

In letters to a college friend who had been in Europe since 1823 Robinson expressed his desire to "tread on *classic* ground," and "for a while to doff Roads & Canals[,] Acqueducts [sic] & bridges . . . to strut on the Boulevards & loll at the Theatre Francois. . . ."[25] But he emphasized that his major objective would be to extend his professional experience, or as he put it:

[20] Moncure Robinson, 31 July 1822, to John Robinson, MRP.

[21] Ibid.; Moncure Robinson, 13 July 1822 to John Robinson, MRP.

[22] Dunaway, 73, 87; Michel Chevalier, *Histoire et description des voies communication aux Etats-Unis,* 2 vols. and plates (Paris: Librairie de Charles Gosselin, 1840–41), 2: 97–98.

[23] "Report of the Commissioner . . . ," *Ninth Annual Report . . . of the Board of Public Works* (Richmond: printed by T. W. White, 1826), 144–49; "Report of the Commissioner . . . ," *Tenth Annual Report . . . Board of Public Works* (Richmond: Shepherd & Pollard, 1826), 305; and receipts and expenses, 1823–25 (uncatalogued) Moncure Robinson box, VaBPW.

[24] Dunaway, *James River and Canal Company,* 72. Moncure Robinson frequently complained of his state's lethargy in internal improvements, e.g., Moncure Robinson, 31 July 1822 to John Robinson, MRP, and Moncure Robinson, 10 February 1827, to John Robinson, MRP.

[25] Moncure Robinson, 11 March 1824, to John Jacquelin Ambler, box 8, AFP; Moncure Robinson, 13 March 1825, to John Jacquelin Ambler, box 7, AFP.

> Instead . . . of Stratford on Avon I should make a pilgrimage . . . to the works of the Duke of Bridgewater. . . . Instead of an excursion in a steam-boat to the Isle of Man I shall probably make one in a post-coach to the factory of those great generators of steam & Steam Engines Messrs. Bolton & Watt. . . .[26]

With the assurance of financial support from his family, Robinson crossed the Atlantic in the spring of 1825 and remained in Europe for two and a half years. He was in England, Wales, and Scotland for at least seven months, although most of the time he lived in Paris.[27] There he attended the Faculty of Science lectures (then held in the Sorbonne) for two winter sessions, including those given by Gay-Lussac, Thenard, De Rozoir, and Le Clerc.[28] The faculty offered public lectures on a variety of topics in the physical sciences.[29] (There is no means of ascertaining which series Robinson attended.) He heard at least one lecture on mechanics at the Conservatoire des Arts et Métiers by Baron Pierre-Charles-François Dupin, an engineer and mathematician who had recently written a book on public works in Britain. Robinson also met with Dupin privately.[30]

Many foreign students attended the public lectures of the

[26] Moncure Robinson, 19 July 1824, to John Jacquelin Ambler, box 7, AFP.

[27] Of preceding American civil engineer visitors, I am aware of Loammi Baldwin, Jr., in 1807–1808 and 1823–24, Canvass White in 1817, and William Strickland and Samuel H. Kneass (by several months) in 1825. These men are noticed in the *Dictionary of American Biography*, 20 vols. (New York: Scribner's, 1927–36) (hereafter *DAB*).

[28] Moncure Robinson, 20 November 1825, 10 February 1826, 21 October 1826, 27 April 1827, to John Robinson, MRP; Moncure Robinson, 19 February 1826, to Conway Robinson; s.v., "Joseph Louis Gay-Lussac," and "Louis Jacques Thenard," in *Dictionary of Scientific Biography*, ed. Charles Coulston Gillespie, 14 vols. (New York: Scribner's, 1970–76) (hereafter *DSB*); Jean Bonnerot, *La Sorbonne: Sa Vie, son rôle, son oeuvre à travers les siècles* (Paris: Les Presses Universitaires de France, 1927), 34.

[29] Maurice Crosland, *The Society of Arcueil: A View of French Science at the Time of Napoleon I* (Cambridge, Mass.: Harvard University Press, 1967), 216–18; Maurice Crosland, *Gay-Lussac: Scientist and Bourgeois* (Cambridge: Cambridge University Press, 1978), 146–49.

[30] 22 January 1826, John Jacquelin Ambler Journal, box 2, AFP; Moncure Robinson, 19 February 1826, to Conway Robinson, MRP; s.v. "Pierre-Charles François Dupin," *DSB*; Frederick B. Artz, *The Development of Technical Education in France, 1500–1850* (Cambridge, Mass.: The Society for the History of Technology and M.I.T. Press, 1966), 216–17.

Robinson states at one point in his Parisian stay that he was "pursuing under private tutors some professional studies"; one of the tutors may have been Dupin. Moncure Robinson, 10 February 1826, to John Robinson, MRP.

Faculty of Science and the Conservatoire.[31] Robinson felt fortunate to be one of them, and realized the benefit of his early training in French.[32] He recognized that there was no comparable opportunity for theoretical studies in the United States, that France was the world's center of engineering education.[33] In fact, it was in 1826, during Robinson's Paris years, that Louis Navier published his lectures on structural analysis and strengths of materials, an event often cited as the birth of modern engineering theory.[34]

In addition to obtaining a formal education in France, Robinson traveled throughout the country to see its public works, including harbors, bridges, waterworks, and steam engines.[35] His primary object was to see canals: a friend with whom he was traveling in France found that Robinson took every opportunity to observe them.[36] When in England with the same friend he walked the towpaths of the Grand Junction and other canals alone to satisfy his curiosity.[37]

It was in fact across the English Channel in Britain that he learned more about the "practice" of civil engineering, which he contrasted with the "theory" of France.[38] At the beginning of his first visit to Britain he reported to his father on the differences between the two countries.

> In practical Mechanics the French must be at least one hundred years behind the English. It is indeed astonishing that in a country so contiguous to one where all the mechanical arts are brought to the highest perfection their contrivances in everything should still be so rude. Here I can travel in no direction, but I come across some fine specimen of [engineering] art. . . .[39]

[31] Crosland, *Gay-Lussac,* 146–49.

[32] Moncure Robinson, 10 February 1826, to John Robinson, MRP.

[33] Artz, *The Development of Technical Education in France,* 266.

[34] Hans Straub, *A History of Civil Engineering,* trans. Erwin Rockwell (Cambridge, Mass.: M.I.T. Press, 1964), 153.

[35] Entries for 22 May 1825, 25 June 1825, 27 June 1825, John Jacquelin Ambler journal, box 2, AFP; Moncure Robinson, 18 August 1826, to John Robinson, MRP.

[36] Entry for 26 June 1825, John Jacquelin Ambler journal, box 2, AFP.

[37] Entries for 5 August 1825, 10 August 1825, 23 August 1825, ibid.; [Moncure Robinson], 16 August 1825, to [Randolf Harrison], Richmond (Va.) *Enquirer,* 25 December 1825. For the identification of the letter see John Robinson, 2 January 1826, to Moncure Robinson, MRP.

[38] Moncure Robinson, 10 February 1827, to John Robinson, MRP.

[39] Moncure Robinson, 22 July 1825, to John Robinson, MRP.

After spending some time in Britain his expressions were even more effusive.

> Much as I heard of the wealth of England I had no conception of it until I came to London. I really am at a loss to know where the poor of this country live. I see every where in England handsome residences & princely dwellings, some cottages but no huts. To Judge superficially one would suppose there was no poverty in the land, for even to live in England seems to require more money than to attain Comfort would anywhere else. . . . And yet with the immense expenditure of England its Capital seems to be accumulating in an unexampled manner. New buildings & Manufactories are springing up every where whilst every new scheme abroad is put into operation with English Capital. What a wonderful country.[40]

The first mention of railroads in Moncure Robinson's correspondence occurs in connection with this trip. He wrote to his father and a friend of his intent to see railroads,[41] and he examined several in his travels through the north of England and Wales.[42] Railroads made a great impression on him, and a letter of his published in the Richmond *Enquirer* late in 1825 included a proposal to use railroads to meet the transportation needs of Virginia. He suggested that the state build a 130-mile railroad with locomotives and inclined planes to connect the canals to be built along the James and Kanawha rivers, so that there would be a modern transportation route from the Ohio River to the tidewater in Virginia.[43]

From his *Enquirer* letter it is clear that Moncure Robinson quickly grasped the feasibility of railroads. He noted the two elements, locomotives and inclined planes, which had changed railroads into their modern form, and he had immediately

[40] Moncure Robinson, 17 September 1825, to John Robinson, MRP.

[41] Moncure Robinson, 22 July 1825, to John Robinson, MRP; [Moncure Robinson], 20 June 1825, to [Randolf Harrison], in Richmond *Enquirer*, 25 December 1825. For the identification of the second letter see John Robinson, 2 January 1826, to Moncure Robinson, MRP.

[42] [Moncure Robinson], 16 August 1825, to [Randolf Harrison], Richmond *Enquirer*, 25 December 1825; *Reports of Moncure Robinson, Esq. & Col. Stephen H. Long, Engineers Appointed by the Canal Commissioners* . . . (Harrisburg, Penna.: printed by Henry Welsh, 1831), 12 (copy in box 8, Allegheny Portage Railroad Reports and Miscellaneous Documents, PaCC).

[43] [Moncure Robinson], 16 August 1825, to [Randolf Harrison], Richmond *Enquirer*, 25 December 1825.

proposed their use in the United States. As a result of this first visit to Britain he changed his professional focus from canals to railroads, and for the rest of his career he never hesitated to urge railroad construction in preference to waterways. To solidify his knowledge of British railroad engineering he returned there in the summer and fall of 1827.[44]

Late in October 1827, Robinson boarded a Liverpool packet and returned to the United States.[45] On arrival he hoped to be employed by Virginia, but that state's legislature showed no intent to start any new internal improvement projects, so he turned to the Pennsylvania state works. He had met some Philadelphians in Paris, at least one of whom had already sent word of Robinson to a prominent relative.[46] From his Parisian acquaintances he had "many letters to persons of the highest standing" in Philadelphia, and with those introductions he intended "to become acquainted with what may be going forward in the way of internal improvements, and with those persons who may have some influence in their management."[47]

In four months he received an appointment from the Pennsylvania Canal commissioners, who assigned him to examine the area between the headwaters of the Schuylkill River, a tributary of the Delaware, and the North Branch of the Susquehanna River. The commissioners directed him to make his surveys "with a view to their [i.e., the two rivers'] connexion by railroad."[48] Apparently they recognized that Robinson had a special understanding of the capabilities of railroads.

[44] Moncure Robinson, 4 September 1827, to John Robinson, MRP. This letter is the only one which dates from that trip. Robinson does not mention railways, but states that "I shall spend some five or six days in visiting the extensive collieries of this place [Sunderland on Wear] & Newcastle. . . ." He must have seen the Stockton and Darlington at that time, although there were many other railroads in the vicinity.

[45] Moncure Robinson, 1 December [1827], to "parents," MRP.

[46] Ibid.; Elihu Chauncey, 30 April 1826, to Nathaniel Chauncey, box 6, Chauncey Family Papers, Yale University Library, New Haven. Two of the friends were Nathaniel Chauncey and Henry Seybert, who are mentioned in Moncure Robinson, "Obituary Notice of Henry Seybert," *Proceedings of the American Philosophical Society* 21 (1883): 248–49.

[47] Moncure Robinson, 9 December 1827, to "parents," MRP.

[48] Jos. M. McIlvaine, 1 April 1828, to Moncure Robinson, MRP. The appointment was made a few days earlier—see entry for 26 March 1827, box 1, Minute Books and Indexes, PaCC.

Within a few weeks Robinson, accompanied by assistants, began his work in the difficult Appalachian ridges of east-central Pennsylvania. His principal assistant was Wirt Robinson, a cousin of his whom the canal commissioners allowed him to appoint, and who was to be his chief aide through much of his civil engineering career.[49] With his party Robinson examined three possible railroad routes from the Schuylkill to the Susquehanna, and also did a rough survey for a proposed canal from the Lehigh River, another tributary of the Delaware, to the Susquehanna along the Nescopec Valley.[50]

His report on the Susquehanna-Delaware surveys shows that he had learned and accepted some British principles of railroad design and construction. In beginning his report he noted that Americans had speculated that railroads could be built more cheaply in the United States than in Britain if wood was used for construction, and that attempts might also be made to build railroads with much steeper grades than were common in Britain. Robinson was willing to agree that the first speculation might come true, but he was certain that the second was wrong. "It is believed," he wrote, ". . . that there is nothing . . . which will authorize a *very material* deviation from the principles adopted by the most distinguished British engineers in the plan and location of such works."[51]

One of those principles was that it was necessary to plan the grade of a railroad with the direction of trade in mind. If equal trade was expected in both directions the road should be as level as possible throughout. If trade was expected in one direction the main restriction was that the road could not be so steep in the return direction so as to hinder the return of empty cars by the motive power. He quoted results from experiments on the Stockton and Darlington Railroad which indicated that this gradation could not exceed eighteen feet to the mile (0.35 percent or about 0.6 degrees).[52]

[49] Moncure Robinson, 6 May 1828, to Joseph McIlvaine, box 1, Surveys & Correspondence, PaCC; Moncure Robinson, 29 June 1828, to John Robinson, MRP.

[50] Moncure Robinson, "Report on the survey of canal and railway routes between the waters of the Delaware and Susquehanna (4 December 1828)," *Hazard's Register of Pennsylvania* (Philadelphia, Penna.), 3 (1829): 54–60; Moncure Robinson, "Report . . . on the Susquehanna and Lehigh Canal (5 January 1829)," *Hazard's Register of Pennsylvania* 3 (1829): 68–72.

[51] Robinson, "Report on the survey of canal and railway routes . . . ," 54–55.

[52] Ibid., 55.

With these principles restricting the grade of railroads, there was little difference between the routes which they could take and the service they could provide and that of canals or turnpikes, Robinson noted. He was referring to the fact that at this date railroads as well as canals and turnpikes depended largely on horses for motive power, and horses could pull about the same amount of weight at the same speed on level canals as near-level railroads. But what had radically changed the applicability of railroads in recent years were locomotives, which had increased the speed of travel, and inclined planes with stationary engines, which allowed railroads to climb grades previously impossible for them.[53] Robinson's recommendation of possible railroad routes was therefore based on four principles:

(1) the routes should cross the mountains at the least possible elevation,
(2) there should be the gentlest grade possible in the direction of most trade, which he expected in this case to be from the Susquehanna River to the south,
(3) routes should be sought which could be used by locomotives, and
(4) intolerable grades should be overcome with inclined planes.[54]

There were echoes of these British ideas throughout the early phases of Moncure Robinson's professional career, even though he was sympathetic to the cheaper wooden construction consonant with American engineering practice.

Only three days after the canal commissioners had received and read his first report on the Susquehanna-Delaware surveys they appointed him engineer on the Allegheny Portage. Their resolution accompanying the appointment asked Robinson to survey the route between the headwaters of the Juniata and Conemaugh rivers and then to give his opinion on whether a railroad, turnpike, "or any other plan" would be the best means of transportation.[55] Behind these words was a political turmoil of more than three years which Julius Rubin discussed in his book *Canal or Railroad?*

With the first wave of enthusiasm for internal improvements

[53] Ibid.

[54] Ibid., 55–56. Many of these ideas were stated earlier in Moncure Robinson's letters to Joseph McIlvaine, box 1, Surveys & Correspondence, PaCC.

[55] Entry for 8 December 1828, box 1, Minute Books and Indexes, PaCC.

in 1825 a canal had been planned from Philadelphia to Pittsburgh which would have needed a four and a half-mile tunnel through the Allegheny summit. Arguments over the feasibility and expense of the tunnel prevented it from being started, although the canal sections in the Allegheny and Susquehanna river systems were begun by the state. The two sections were useless for east-west trade without some sort of mountain connection, however, and by the time Moncure Robinson was appointed as engineer Pennsylvania merchants and politicians were involved in a furious debate over what was to be done.[56] The canal commissioners turned to Robinson as the engineer who had the technical skill to provide the answer.

He took the job with no thought that anything but a railroad would answer the purpose,[57] and exhilarated by the challenge of being in charge of a significant and well-publicized project.[58] In a letter to his father he explained that:

> I am to push my explorations into every glen and nook of the Allegheny through which it may be possible to creep with a railroad. Should I succeed in finding a route favorable enough to be adopted and executed I shall be much pleased, as it will afford me an opportunity of making myself known in a department of my profession, in which few engineers have as yet embarked.[59]

Robinson worked on the Allegheny Portage survey with a party of six assistants from about April until August of 1829.[60] In his report filed three months later he reiterated the principles stated in the Susquehanna-Delaware report and based his recommendations on them. He suggested a railroad using locomotive power on the most level segment, water and steam power on several inclined planes which would overcome steep grades, and a one-mile tunnel to reduce the height of the summit.[61]

[56] Rubin, *Canal or Railroad?,* 20–56.

[57] Moncure Robinson, 15 December 1828, to John Robinson, MRP.

[58] Moncure Robinson, 5 April 1829, to Conway Robinson, MRP; Moncure Robinson, 5 April 1829, to John White, Adelman Collection, HSP.

[59] Moncure Robinson, 6 April 1829, to John Robinson, MRP; item in "List of Persons, Employed by the Canal Commissioners, Allegheny Portage & Rail Road," box 2, General Correspondence and Surveys, PaCC.

[60] Ibid., Moncure Robinson, 18 August 1829, to John Robinson, MRP.

[61] "Report of Moncure Robinson, Principal Engineer upon the Allegheny Portage (21 November 1829)," *Pennsylvania House Journal* (Harrisburg), 1829–30, II, doc.

Unfortunately, this survey was as difficult for the Pennsylvania assembly to accept as that of 1825: no action was taken on it, and in March 1830 another survey was authorized. A board of three engineers appointed by the canal commissioners was instructed "to take into view a portage by means of a road so graded as to admit of its being adapted either to a McAdamized turnpike or a railroad."[62] Moncure Robinson was appointed to the board along with Colonel Stephen H. Long and (after Major David B. Douglass declined the appointment) Major John Wilson of the Army's Engineers.[63] He spent little time with the other two engineers as they carried out their survey, because he was content with his previous work.[64]

Long and Wilson submitted a report in December of 1830 in which they agreed with Robinson's conclusion of a year earlier that a railroad would be the best means of crossing the summit, but objected to his tunnel and to the steepness of his inclined planes.[65] Colonel Long later outlined the details of their proposal in a separate report.[66]

Robinson filed a rebuttal in March which was a blistering attack on Long's engineering principles, especially those which concerned the inclined planes. He examined current engineering thought, then considered railroads in England and Wales, and concluded that "if the theory of the inclined plane furnishes nothing to justify the limit of acclivity assumed by Lt. Col. Long, neither does the practice of British engineers."[67] He supported himself with letters from Horatio Allen and John Jervis, the two engineers on the only American railroad with a number of inclined planes.[68]

138. (Reprinted in *Transactions of the American Society of Civil Engineers* 15 [1886]: 183–202, to which page references will be made.)

[62] Rubin, *Canal or Railroad?*, 58; 27 March 1830, "Journal of the Canal Commissioners," *Pennsylvania House Journal*, 1830–31, II, doc. 82.

[63] Ibid.; F. R. Shunk, 29 March 1830, to Moncure Robinson, MRP; Moncure Robinson, 2 April 1830, to Francis R. Shunk, box 1, Reports and Miscellaneous Documents, Allegheny Portage Railroad, PaCC.

[64] Moncure Robinson, 6 May 1831, to Board of Canal Commissioners, box 1, Reports and Miscellaneous Documents, Allegheny Portage Railroad, PaCC.

[65] John Wilson and S. H. Long, "Report of the Engineers appointed to survey the Allegheny Portage (18 December 1830)," *Pennsylvania House Journal*, 1830–31, II, doc. 27.

[66] *Reports of Moncure Robinson, Esq. & Col. Stephen H. Long, Engineers . . .*, 19–23.

[67] Ibid., 11.

[68] Ibid., 13.

Robinson's defense of his own work was well taken by the Board of Canal Commissioners and although Sylvester Welch was appointed to take charge of the construction of the Allegheny Portage Railroad, Robinson was requested to be available for consultation.[69] Welch wrote to him at least once concerning an engineering question, and in the same letter stated that it was Robinson's survey of 1829 that he was using as his guide rather than Long's.[70] As a result, although Welch did not adopt the tunnel and altered the design of the planes somewhat, the Portage was built essentially as Robinson had envisioned it.[71]

The Pennsylvania assembly's rejection of his original recommendations made Moncure Robinson turn for employment from state to private enterprise. That change had been coming for some time, as he had been involved with some railroad promotional activities almost since his return from Europe. In fact, for the entire decade beginning with the summer of 1829 he was constantly connected with non-public railroad projects, and seldom performed professional services for a state. He took on several commissions concurrently in these years which makes it impossible to discuss them chronologically. It is, however, convenient to divide them into his "early" railroads—those on which he was appointed chief engineer up to 1833, and his "late" railroads—those on which his appointment was after 1833. This division illuminates Robinson's shift from the British engineering principles upon which he relied in his reports to the canal commissioners and his rebuttal to Long, to more typically American approaches.

The first of his early railroads was in Virginia. For a number of years Virginians had been interested in building a railroad from the Midlothian coal mines southwest of Richmond to the falls of the James River, a distance of about thirteen miles. Benjamin Henry Latrobe had suggested that a railroad be built there,[72] and the state engineer of Virginia concluded after a survey in 1828 that "a railway from the coal-pits to

[69] Entry for 30 March 1831, "Journal of the Board of Canal Commissioners," *Pennsylvania House Journal,* 1831–32, II, doc. 99.

[70] Sylvester Welch, 18 April 1831, to Moncure Robinson, MRP.

[71] Chevalier, *Histoire et description,* 1: 398.

[72] Stapleton, ed., *The Engineering Drawings of Benjamin Henry Latrobe,* 28, 139, 140.

the James River is perfectly practicable."[73] Robinson was involved in the agitation for chartering a railroad company which followed the state survey but the first attempt fell through, partly due to his absence in Pennsylvania.[74] In the following spring the effort was successful, and in July 1829, Robinson was appointed the engineer of the Chesterfield Railroad.[75] That fall he laid out the route of the railroad,[76] and after construction began gave his cousin Wirt direction of the work.[77] He returned to supervise construction in the winter of 1830–31, and the railroad was completed the spring of the latter year.[78] Robinson did not install steam power on the Chesterfield, perhaps because the capital investment was very small.[79] Mules provided the motive power, and the one inclined plane was "self-acting," that is, the weight of loaded cars was used to pull up empty cars.[80]

The Chesterfield was both a technical and economic success. Although there was at least one early accident on the inclined plane,[81] the railroad operated successfully until competition with a parallel and more modern railroad put it out of business in the early 1850s. Until that time it paid excellent dividends to the stockholders.[82]

Robinson became involved with the promotion of the Danville and Pottsville Railroad through the Susquehanna–Delaware surveys.[83] The canal commissioners ordered him to make a "topographical plan" showing possible railroad routes

[73] Claudius Crozet, "Report on the Rail-Way from the Coal-Pits to James River," *Twelfth Annual Report of . . . the Board of Public Works* . . . (Richmond: Samuel Shepher & Co., 1828), 22.

[74] Francis Earl Lutz, *Chesterfield: An Old Virginia County* (Richmond: William Byrd Press, Inc., 1954), 181; Conway Robinson, 21 March 1829, to Moncure Robinson (typescript), MRP. Other sources state that the charter attempt was successful in 1828, but the letters in the MRP indicate that it was the following year.

[75] Lutz, *Chesterfield,* 181–82; Cary Robinson, 21 July 1829, to Moncure Robinson, MRP.

[76] Moncure Robinson, 18 October 1829, to Patrick Durkin et al., MRP.

[77] Elizabeth Dabney Coleman, "Forerunner of Virginia's First Railway," *Virginia Cavalcade* 4.3 (Winter 1954): 5.

[78] Ibid.; Richmond *Enquirer,* 21 June 1831.

[79] Chevalier, *Histoire et description,* 2: 488.

[80] Ibid.; Richmond *Enquirer,* 21 June 1831.

[81] John Robinson, 17 July 1831, to Moncure Robinson, MRP.

[82] Chevalier, *Histoire et description,* 2: 488; Coleman, "Forerunner," 6–7.

[83] Moncure Robinson, 6 May 1828, and 28 May 1828, to Joseph McIlvaine, box 1, Surveys and Correspondence, PaCC; *Hazard's Register of Pennsylvania* 16 (1835): 402.

between the two towns.[84] He also attended a meeting of Philadelphians interested in the railroad in February 1829, when he presented his "report, plans, and profiles, accompanied by satisfactory details and explanations."[85]

It was almost two years later before investors showed enough interest in the railroad for the stock subscription books to be opened.[86] Even then there was insufficient investment, and it took a special appeal for funds from Stephen Girard, the Philadelphia financier who owned a large tract of land on the route of the projected railway, to gather enough capital.[87] In May 1831, Moncure Robinson was asked to begin final location of the railroad,[88] and in July and August he directed a corps of assistants in that task.[89]

Although at first he shared the duties of engineer with F. W. Rawle,[90] Robinson later became chief engineer and directed construction up to the opening of the railroad in 1835.[91] He planned and executed six inclined planes for the Danville and Pottsville, each of which were intended to be self-acting. Since, however, loaded cars might not always be available to draw up empty ones, he designed special water-tank cars which were filled at the top of the planes to provide ballast.[92] This ingenious system was supplemented at the longest and steepest plane (number five, the Mahanoy Plane) by a steam engine of ninety-five horsepower.[93] The Danville and Pottsville Railroad was partially completed in 1835, but the difficult central portion of about fourteen miles was never begun. The eastern

[84] Moncure Robinson, 19 February 1829, to Joseph McIlvaine, box 1, Surveys & Correspondence, PaCC.

[85] *Hazard's Register of Pennsylvania* 6 (1830): 378.

[86] *The Miners' Journal* (Pottsville, Penna.), 1 January 1831.

[87] Daniel Montgomery, 23 March 1831, to Stephen Girard, reel 117, II, Stephen Girard Papers, American Philosophical Society, Philadelphia, Penna. (hereafter SGP); Stephen Girard, 25 March 1831, to Daniel Montgomery, Society Collection, HSP; Daniel Montgomery, 28 March 1831, to Stephen Girard, reel 117, II, SGP.

[88] Minutes of the Danville and Pottsville Rail Road Company, 5 May 1831, HML.

[89] *The Miners' Journal,* 2 July 1831, 6 August 1831, 20 August 1831.

[90] D and P Minutes, 5 May 1831, 11 May 1831; *The Miners' Journal,* 28 May 1831.

[91] Rawle was fired in 1832, and Robinson resigned early in 1836: 21 May 1832, 13 January 1836, D and P Minutes.

[92] Chevalier, *Histoire et description,* 2: 511. There is a translation of some of Chevalier's description of the Danville and Pottsville, and a reproduction of some of the accompanying plates in John N. Hoffman, *Girard Estate Coal Lands in Pennsylvania, 1801–1804* (Washington, D.C.: Smithsonian Institution, 1972), 69–76.

[93] Chevalier, *Histoire et description,* 2: 509.

half was essentially abandoned in 1838 because output from the neighboring coal mines produced insufficient revenue to keep it operating. Moncure Robinson's magnificent inclined planes, which were all on that section, had a very short life.[94]

Another enterprise in the Pottsville vicinity with which Robinson was associated was the Mount Carbon Railroad. It connected landings at the uppermost basin of the Schuylkill canal with coal mines about seven miles away. Robinson had some role in getting a charter for the railroad in the spring of 1829,[95] but it was not until August that he was chosen to survey the route.[96] Accompanied by members of the Mount Carbon's board of managers he made a rapid survey early in September,[97] but declined the appointment as engineer.[98] The man he suggested for the position was appointed, and when that engineer resigned some months later, Robinson was asked "to direct the [new engineer] in the plan of proceeding to the completion of the Rail Road."[99] Robinson's advice had been followed earlier concerning the kind of rails and spikes to be used for the track.[100] Completed in 1831, the Mount Carbon Railroad was an important coal-carrying route for many years.[101]

The Little Schuylkill Railroad was the most quickly built of Robinson's early railroads in the Schuylkill County anthracite district. He was solicited for the position of chief engineer of the Little Schuylkill in the fall of 1829,[102] but not actually appointed until March 1830.[103] In June he arrived on the site and began the work.[104] The terrain was not difficult, as the route followed the Little Schuylkill River for twenty-

[94] Ibid., 2: 521–22; Hoffman, *Girard Estate,* 65–66.

[95] Moncure Robinson, 5 April 1829, to John White, Adelman Collection, HSP.

[96] Minutes of the Board of Managers of the Mount Carbon Railroad Company, 29 August 1829, HML (hereafter Mt. Carbon Minutes).

[97] Ibid., 17 September 1829, 22 September 1829.

[98] Ibid., 17 September 1829.

[99] Ibid., 10 June 1830.

[100] John White, 10 December 1829, to Moncure Robinson, MRP.

[101] Henry V. Poor, *History of the Railroads and Canals of the United States of America* (New York: John H. Schultz & Co., 1860), 1: 462.

[102] Thos. Biddle per Edward C. Biddle, 20 November 1829, to Moncure Robinson, MRP.

[103] Jay V. Hare, *History of the Reading* (Philadelphia: John Henry Strock, 1966, reprint), 225.

[104] Moncure Robinson, 29 June 1830, to John Robinson, MRP.

one miles from the coal region around Tamaqua to the Schuylkill Canal at Port Clinton, but Robinson had two problems to contend with. One, the necessity of keeping costs down, was imposed by the small amount of capital accumulated by the company; the other, the desirability of building the road so as to be strong enough to use locomotives, Robinson placed on himself.[105] He solved the problems so well that the Little Schuylkill Railroad was opened in 1831,[106] and two years later imported English locomotives were successfully used on it.[107]

Contemporary opinion of Robinson's skill in building the Little Schuylkill was best expressed by the President of the Pennsylvania Canal Commissioners, who, in declining an invitation to come to the opening of the railroad, stated: "I regret the circumstance exceedingly, as . . . [the] invitation would have afforded me an opportunity of perfecting my knowledge of Rail Roads, by inspecting one built under the direction of the first Master in the United States."[108]

The first long interregional railroad of which Moncure Robinson was chief engineer was the Petersburg, which extended sixty miles from Petersburg, Virginia, to near Weldon, North Carolina. This railroad was intended to connect the rich agricultural area of upper North Carolina with seaport facilities at Petersburg.[109] Robinson was asked to make a survey of the proposed route in the fall of 1829,[110] and carried out the work in February 1830.[111] In his survey report he referred again to the principle of descending tonnage; that is, that in the direction of most trade the grade of a railroad should be level or "downhill." He noted that it was almost impossible to adhere to that principle in this case, due to the numerous "streams and ridges" which the railroad had to cross, and that therefore only locomotive power could overcome that disadvantage.[112] After Robinson was appointed chief

[105] Chevalier, *Histoire et description,* 2: 522.
[106] *Hazard's Register of Pennsylvania* 8 (1831): 361.
[107] Chevalier, *Histoire et description,* 2: 523–24.
[108] James Clarke, 14 November 1831, to Moncure Robinson, MRP.
[109] Chevalier, *Histoire et description,* 2: 419–20.
[110] Moncure Robinson, 18 October 1829, to Patrick Durkin et al., MRP.
[111] *Report of the Engineer, on a Survey of a Rail-Road From Petersburg to the Roanoke* ([Petersburg]: n.p., 1830), 3.
[112] Ibid., 4–5.

engineer the company adopted his recommendations, and when the first section was opened late in 1832 there were already two English locomotives on hand to operate it.[113] Other English locomotives were imported in later years.[114]

The last of Moncure Robinson's early railroads was the Winchester and Potomac, which connected Winchester, Virginia, with the Baltimore and Ohio Railroad at Harpers Ferry. Robinson was chosen chief engineer early in 1833,[115] and began letting contracts for the thirty-two-mile route in the fall of the year.[116] Apparently he was the engineer during most of the construction of the road, although he must have left an assistant in charge much of the time because of the demands of his other railroads.[117] The Winchester and Potomac was opened for operation in 1836,[118] and the motive power was provided by four English locomotives.[119]

In planning and locating this group of early railroads Moncure Robinson adhered to the British principles which he had so clearly stated in his report on the Susquehanna-Delaware surveys. He kept grades as low as possible, recommended locomotives where they could be used to advantage, and used inclined planes to overcome large changes in elevation. He also attempted to follow British practice in the construction of the railroads and the equipment used in their operation.

[113] Moncure Robinson, 6 December 1832, to J. Brown, Jr., *Seventeenth Annual Report . . . Board of Public Works* (Richmond: Samuel Shepherd and Co., 1833), 43–47; Robert R. Brown, "Pioneer Locomotives of North America," *Railway and Locomotive Historical Society Bulletin* 101 (October 1959): 50–52.

[114] Brown, "Pioneer Locomotives," loc. cit.

[115] Moncure Robinson, 21 March 1833, to Conway Robinson, MRP.

[116] *Niles' Weekly Register* (Baltimore, Md.), 21 September 1833.

[117] Chevalier, *Histoire et description,* 2: 37. When Chevalier visited the Winchester and Potomac in 1834 Moncure Robinson was chief engineer, but there are documents which indicate that he resigned at one point (minute of the President and Directors of the Winchester and Potomac Railroad Company, 28 June 1834, MRP; Wm. H. Morell, 7 July 1834, to Moncure Robinson, MRP). The only indication thereafter that Moncure was associated with the railroad comes in three letters dated from Winchester: Moncure Robinson, 31 January 1835, to Board of Public Works, (uncatalogued) Moncure Robinson box, VaBPW; Moncure Robinson, 31 January 1835, to the General Assembly of Virginia, *Nineteenth Annual Report of . . . the Board of Public Works* (Richmond: Samuel Shepherd, 1835), 411–13; Moncure Robinson, 2 February 1835, to James Brown, Jr., (uncatalogued) Moncure Robinson box, VaBPW. Robinson must have had an assistant in charge of the work because the Danville and Pottsville, Philadelphia and Reading, and R, F and P were under his supervision in 1834–35.

[118] *Niles' Weekly Register,* 26 March 1836.

[119] Brown, "Pioneer Locomotives," 76.

Many British colliery railroads were constructed with a wooden superstructure, that is, with wooden cross-ties, or sleepers, laid a few feet apart on the roadbed, and wooden rails laid parallel on top four feet eight and a half inches (the modern standard gauge) apart. On some of these wooden railroads strips of malleable plate iron were spiked to the top of the rail to prevent excessive and uneven wear on the rail.[120] A major difficulty with this sort of superstructure was that the horses' hooves (and shoes) tended to wear through the wooden cross-ties.[121] This problem was solved by replacing the cross-ties with rows of stone blocks under each rail, the blocks being partially sunk into the ground. While that left a clear horsepath, the stone was expensive to purchase and transport, and was less capable of keeping the rails to proper gauge.[122] On the Stockton and Darlington, and Liverpool and Manchester, the stone blocks were combined with rails made completely of cast or wrought iron ("edge rails") to make what was believed to be a "permanent way."[123]

American civil engineers were therefore able to draw upon more than one British tradition when they planned the rail and track for American railroads. Impressed by the solidity and apparent permanence of the Liverpool and Manchester example a number of early American railroads, such as the Boston and Lowell, and Columbia and Philadelphia, had tracks built with stone blocks and iron edge rails.[124] Some other railroads, such as the New Castle and Frenchtown, and New York and Harlem, also had stone block sleepers, but used wooden rails plated with iron.[125] Both the edge rails and the plate iron were imported from England.[126]

[120] Nicholas Wood, *A Practical Treatise on Railroads,* 3rd ed. (London: Longman, Orme, Brown, Green, & Longmans, 1838), 20–23.

[121] Ibid., 20; Rolt, *George and Robert Stephenson,* 75.

[122] Wood, *Practical Treatise,* 20, 150–51; Rolt, *George and Robert Stephenson,* 75.

[123] Von Oeynhausen and von Dechen, *Railways in England,* 11–12, 24–25, 44–45; Carlson, *Liverpool and Manchester,* 198.

[124] "Columbia and Philadelphia Railroad. Report of William B. Mitchell, Superintendent," *Pennsylvania House Journal,* 1832–33, II, doc. 9; J. Knight and Benj. H. Latrobe, *Report Upon the Plan of Construction of Several of the Principal Rail Roads in the Northern and Middle States* (Baltimore: Lucas and Deaver, 1838), 11–12.

[125] Knight and Latrobe, *Report,* 4; William F. Holmes, "The New Castle and Frenchtown Turnpike and Railroad Company, 1809–1838: Part II," *Delaware History* 10 (October 1962): 174–75.

[126] "Columbia and Philadelphia Railroad. John Barber, Superintendent," *Pennsylvania House Journal,* 1831–32, II, doc. 22; 23 November 1830, Minute Book of

Moncure Robinson was fully aware of the expense of constructing a track with stone blocks and edge rails, even if it was believed to be permanent. He was therefore sympathetic to the use of as much wood as possible, especially since wood was cheaper in the United States than in Britain.[127] In recommending the kind of track to be used on the Allegheny Portage in 1829 he stated that:

> In the superstructure of the railroad more economy may be exercised than in grading or roadway formation. The cheapness of timber, and the facility with which it may be renewed, will recommend, in the first instance at any rate, wooden rails plated with iron bars in preference to rails of malleable or cast-iron. A superstructure of this description is recommended by the further consideration that a less expensive description of iron may be made use of for plating wooden rails than would be required for rails entirely of metal.[128]

All his early railroads had track of this type. The wooden cross-ties were usually of oak, and the wooden rails normally of southern yellow pine. The plate rail was usually about two inches wide and a half-inch thick, and was imported from England. The plate rail was spiked on to the wooden rail, which in turn was held in notches in the cross-tie with wooden wedges.[129]

In the days before wood preservatives, deterioration of the wooden part of the track was a major problem. Locust, cedar, oak, and pine were preferred woods because of their rot-resistant qualities, but they were not always available,[130] and

the New Castle and Frenchtown Turnpike and Railroad Company, microfilm, Acc. 722, HML.

[127] Robinson, "Report on the Survey" (n. 50 above), 54; 17 October 1831, W. D. Lewis diary, Historical Society of Delaware, Wilmington. See the excellent discussion of Americans' use of wood for railroads by John H. White, Jr., "Tracks and Timber," *IA: The Journal of the Society for Industrial Archeology* 2 (1976): 35–46.

[128] "Report of Moncure Robinson" (n. 61 above, p. 193 in *TASCE* reprint).

[129] Moncure Robinson, 6 December 1832, to John Brown, Jr. (n. 113 above); Chevalier, *Histoire et description,* 2: 37, 419–21, 520–21; Moncure Robinson, 2 October [1830], to Wirt Robinson, Acc. 1520, HML; Moncure Robinson, 5 October 1830, to Canvass White, Canvass White Papers, John M. Olin Library, Cornell University, Ithaca, N.Y.; John White, 12 December 1829, to Moncure Robinson, Adelman Collection, HSP; Moncure Robinson, 1 June 1831, to Stephen Girard, reel 118, II, SGP; Stephen Girard, 18 November 1831, to Moncure Robinson, reel 128, III, SGP.

[130] Chevalier, *Histoire et description,* 2: 37; Knight and Latrobe, *Report.*

even they deteriorated when constantly exposed to moisture. Robinson recognized this situation,[131] but with his early railroads his major method of dealing with it seems to have been only the use of rot-resistant woods.[132] He did believe in some use of gravel or stones to keep the track in place,[133] but apparently did not recognize that gravel ballasting also promoted drainage, thus keeping the wood drier and reducing rot. In order to keep down costs on the Petersburg Railroad Robinson virtually dispensed with broken stone,[134] and the superstructure which he thought would last ten years had to be replaced in less than half that time.[135]

Moncure Robinson's railroads did not, however, suffer from the problems of those which had stone blocks to hold their rails. In areas where there were long periods of freezing weather the blocks were heaved by the action of the frost, and when a thaw set in they tended to spread apart.[136] More serious, though, was the inelastic nature of the stone and iron combination when used with locomotives. The weight and rhythmic action of locomotives wore heavily on the rails and stone blocks because they did not "give."[137] Faced with such problems engineers began to realize that tracks with stone sleepers, which were believed to be permanent, were actually defective. Moncure Robinson, in choosing what appeared to be the "penny-wise but pound-foolish" alternative from the British heritage of railroad technology, had actually chosen better than some other American railroad engineers.

Following the British example, Americans also made inclined planes part of many railroads in the United States. Inclined planes may be looked at as a solution to the conflict between keeping to a low grade and making the distance traveled as short as possible, but the major reason for their

[131] "Report of Moncure Robinson" (n. 61 above, pp. 193–94 in *TASCE* reprint).

[132] E.g., Chevalier, *Histoire et description,* 2: 37.

[133] Moncure Robinson, 2 October [1830], to Wirt Robinson, HML.

[134] Chevalier, *Histoire et description,* 2: 420; Moncure Robinson, 6 December 1832, to James Brown, Jr. (n. 133 above); *Report of the Engineer* (n. 111 above), 6.

[135] Chevalier, *Histoire et description,* 2: 422.

[136] Knight and Latrobe, *Report,* 11, 52; Darwin H. Stapleton, "Borrowed Technology on the Allegheny Portage Railroad," *Canal Currents* (Bulletin of the Pennsylvania Canal Society) (Spring 1973): 6; David Stevenson, *Sketch of the Civil Engineering of North America* (London: John Weale, 1838), 239.

[137] Knight and Latrobe, *Report,* 46; Stevenson, *Sketch,* 251.

use in the United States was engineers' initial unquestioning adherence to British principles. George Escol Sellers, who knew many of the early American railroad engineers put it this way:

> We are all naturally disposed to follow leads. The great canals of the world made tunneling a necessity; much talent and skill was expended on them. To the tunnels the master spirits of the Liverpool and Manchester added the inclined plane; our following was natural, and Pennsylvania was not alone in doing so on her Columbia and her Portage Railroad. The Charleston & Hamburg, the Lawrenceburg & Indianapolis and others might be cited, all of which planes have been superseded by gradual grades, worked by the ordinary locomotive.[138]

Moncure Robinson was guided by British experience when he installed inclined planes on the Chesterfield, the Petersburg, and the Danville and Pottsville railroads. Robinson later, however, recognized that planes could usually be avoided. In 1835 when he was asked to confer with Jonathan Knight and Benjamin Wright on the proposed location of the New York and Erie Railroad, he concurred that there were "general objections to inclined planes on a railroad, on which the rapid transit of passengers and merchandise is desirable."[139]

Robinson apparently preferred English locomotives for the early railroads which he designed for locomotives. Although American manufacturers were producing locomotives early in the 1830s,[140] on the Little Schuylkill, the Petersburg, and the Winchester and Potomac, the first locomotives were of English build.[141] Robinson's role in having the locomotives ordered from England is unknown, but he undoubtedly exercised influence in the decision, since the companies depended upon him for technical advice. Moreover, his assistant Wirt Robinson was sent to Britain by the Petersburg Railroad

[138] Ferguson, ed., *Early Engineering Reminiscences,* 150–51; cf. W. Milnor Roberts, "Reminiscences . . . with Especial Reference to Steep Inclines," *Transactions of the American Society of Civil Engineers* 7 (1878): 202.

[139] *Report of Moncure Robinson, . . . Jonathan Knight, . . . and Benjamin Wright . . . Upon the Plan of the New-York and Erie Railroad* (New York: George P. Scott and Co., 1835), 8.

[140] John H. White, Jr., *American Locomotives* (Baltimore: Johns Hopkins Press, 1968), 13, 33.

[141] Brown, "Pioneer Locomotives," 38–39, 50–52, 76.

to purchase its locomotives,[142] and while there probably ordered those for the Little Schuylkill as well.[143]

Wirt's visit to Britain is interesting because it seems to have had more importance than just placing equipment orders. He went to Britain in the spring of 1831, and returned in January 1833,[144] after having visited England, Scotland, and Wales.[145] He seems to have taken a special interest in inclined planes,[146] probably because his mentor wanted information on British machinery before he built the planes for the Danville and Pottsville.[147] Moncure Robinson clearly regarded Wirt's visit as an important one, as he expected him to be "deriving all the benefits to be attained by an extensive and minute examination of all the English Improvements."[148] In a letter to John Jervis, then chief engineer of the Mohawk and Hudson Railroad, he wrote that:

> It occurs to me that there may be some facts or plans of machinery, or other matters which you might like to have from England. If there are I could get at any thing of the kind at this time with great readiness for you through a very industrious assistant & near relative Mr. Wirt Robinson who has been there for some time and who has I believe very free access to every thing in the way of our profession.[149]

[142] "Wirt Robinson for services in England," in "Statement shewing the disbursements in the Engineer Department, in the Construction of the Petersburg Rail Road . . . [1830–1833]," box 35, VaBPW. Cf. Ferguson, ed., *Early Engineering Reminiscences,* 134. Sellers refers to the locomotives as ordered for the Richmond, Fredericksburg and Potomac Railroad, an unlikely event since it was not yet incorporated.

[143] The locomotives for the Little Schuylkill were made by the same manufacturer as those for the Petersburg. Brown, "Pioneer Locomotives," 38–39, 50–52.

[144] *The Miners' Journal,* 28 May 1831, stated that "a gentleman, related to Mr. [Moncure] Robinson, has taken a voyage to England." George Escol Sellers, who remembered having passage home on the same ship with Wirt Robinson, returned in January 1833: Ferguson, ed., *Early Engineering Reminiscences,* 108, 134.

[145] Wirt Robinson notebook [1831–32], MRP; Wm. Armstrong, 26 December 1831, to Wirt Robinson, and Wm. Armstrong, 14 February 1832, to Wirt Robinson, Acc. 1520, HML.

[146] Ibid. (all of n. 145 above); Wm. Armstrong, 9 December 1831, to Wirt Robinson, Acc. 1520, HML.

[147] *The Miners' Journal,* 28 May 1831; Moncure Robinson, 5 May 1831, to D. Montgomery et al., 5 May 1831, D and P Minutes. In August 1832, while Wirt was abroad, Moncure Robinson was authorized by the Danville and Pottsville Railroad "to procure from England such portion of the machinery of the Rail Road as may seem to him expedient" (8 August 1832, D and P Minutes), but there is no evidence that any machinery was imported.

[148] Moncure Robinson, 10 June 1832, to Conway Robinson, MRP.

[149] Moncure Robinson, 4 June 1832, to John B. Jervis, box 30, Jervis Papers, Jervis Public Library, Rome, N.Y.

Wirt Robinson's visit to Britain emphasizes Moncure Robinson's reliance on British railroad technology during the construction of his early railroads. But in building his later railroads Moncure Robinson developed a blend of British technology and American values which indicate maturity as a civil engineer. These railroads clearly show the beginning of an American railroad technology.

In 1833 Robinson was involved in the agitation for a railroad from Richmond to Fredericksburg, Virginia, and the Potomac Creek north of Fredericksburg.[150] When the company was organized in 1834 as the Richmond, Fredericksburg, and Potomac Railroad (R, F and P), he was selected as chief engineer. Some of the route was under contract by the end of that year, and the entire fifty-mile route was opened in January 1837, when there were six English locomotives in service.[151] This railroad cut through a rolling country crossed by several rivers, and was destined to become a major link in the rail network of the eastern seaboard.

Moncure Robinson was the logical choice as chief engineer of the Richmond and Petersburg Railroad, a twenty-one-mile line which connected his other major efforts in Virginia, the Petersburg, and the R, F and P. He surveyed the route for the interested parties in 1835, and concluded that a railroad could be built between the two cities, but that it would involve two expensive bridges: one over the Appomattox River at Petersburg, and another over the James River at Richmond.[152] The company was incorporated early in 1836,[153] and Robinson was chosen chief engineer soon after.[154] Work began quickly, and the entire route was opened for travel using English locomotives in May 1838.[155] The bridge which he constructed over the James River was regarded as a major

[150] Moncure Robinson, 31 July 1833, to J. H. Pleasants, *American Railroad Journal* (New York), 2 (1833): 561. The Potomac *Creek* is a tributary of the Potomac *River.*

[151] John B. Mordecai, *A Brief History of the Richmond, Fredericksburg and Potomac Railroad* (Richmond: Old Dominion Press, 1941), 5, 8–9.

[152] Moncure Robinson, *Report of the Chief Engineer on the Richmond and Petersburg Rail Road* (Richmond: T. W. White, 1836).

[153] *Charter of the Richmond and Petersburg Rail Road Company* (Richmond: T. W. White, 1836), 11.

[154] *Second Meeting of the Stockholders* [of the Richmond and Petersburg Rail Road] (Richmond: n.p., 1837), 33.

[155] *Proceedings of the Stockholders in the Richmond and Petersburg Rail Road Company* . . . (Richmond: Thomas W. White, 1839), 61; Chevalier, *Histoire et description,* 2: 418; Brown, "Pioneer Locomotives," 62.

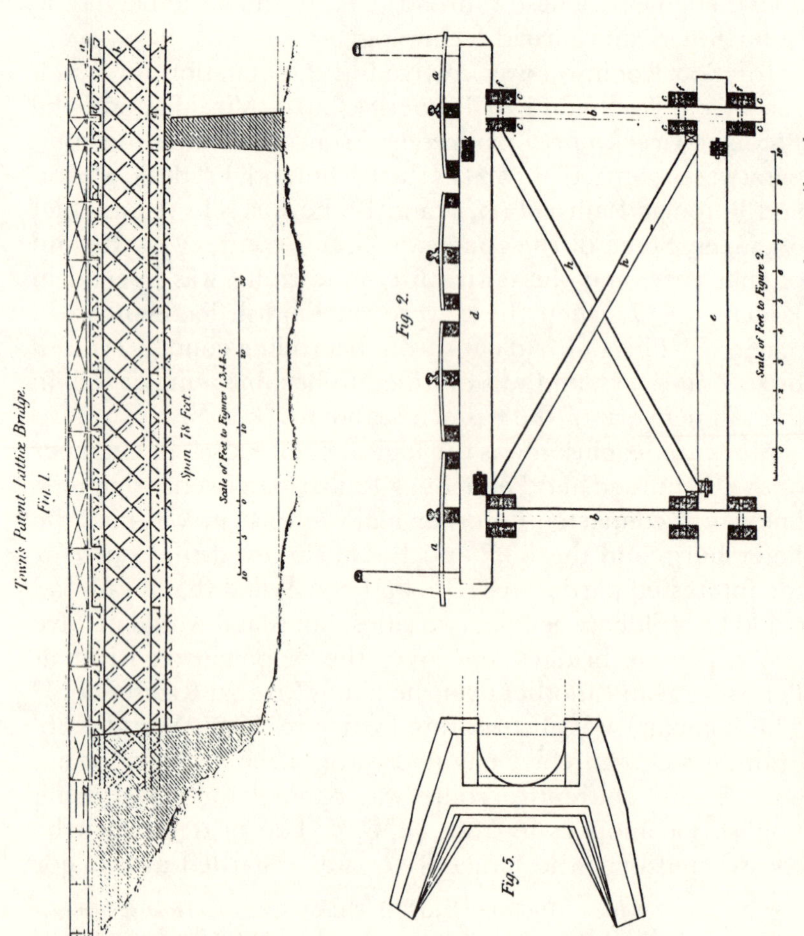

Fig. 10. Bridges of Town's design were erected on Robinson's later railroads.

engineering accomplishment by Michel Chevalier, a French engineer.[156] It was the longest bridge of wooden lattice trusses (of the Town type) ever built, having nineteen spans totalling 2,820 feet.[157] (See fig. 10.)

The third of his three later railroads, the Philadelphia and Reading, required the most of Moncure Robinson's engineering skill. This railroad was chartered in 1833 to run between the two Pennsylvania cities, and was later allowed to extend its route to the anthracite coal district at Pottsville.[158] As it was expected to carry a heavy coal traffic Robinson adhered to strict rules in locating the route so that it would be easily negotiable by locomotives. These rules are clearly reminiscent of those based on English experience which he enunciated in his 1828 report on the Susquehanna-Delaware surveys: (1) no grade in the direction of trade should be ascending, (2) no other grade should exceed eighteen feet to the mile, and (3) the shortest radius of curvature should be 818 feet.[159] The narrow, sinuous valley of the Schuylkill which the railroad followed, and which already contained turnpikes and a canal, was not the easiest situation in which to adhere to any of these rules. The railroad required three tunnels, at least one masonry bridge, and several large, wooden lattice-truss bridges.[160]

He had conducted a preliminary survey of the line in the summer of 1834, but Robinson was not chosen chief engineer of the Philadelphia and Reading Railroad until March 1835. He then rapidly assembled a large corps of engineers to assist in location and construction.[161] The work began late in 1835, with the section from Philadelphia to Reading ready in 1838,

[156] Chevalier, *Histoire et description,* 2: 417, 571–75.

[157] Carl W. Condit, *American Building: Materials and Techniques from the Beginning of Colonial Settlements to the Present* (Chicago: University of Chicago Press, 1968), 59.

[158] Hare, *History of the Reading,* 2, 7.

[159] Osborne, *Professional Biography,* 25–26.

[160] Ibid., 30–31, 33–34; Moncure Robinson, *Report on the Philadelphia and Reading Railroad* (Philadelphia: n.p., 1834); Stevenson, *Sketch,* 231, 234; Knight and Latrobe, *Report,* 23; notes on the Philadelphia and Reading Rail Road [1835–38], W. H. Wilson Notebook, microfilm in private collection; 11 September 1835, 8 October 1835, 2 January 1836, W. M. C. Fairfax diary, VHS.

[161] Hare, *History of the Reading,* 3; 15 March 1835, 30 March 1835, 5 June 1835, W. M. C. Fairfax diary, VHS; W. Hasell Wilson, *Reminiscences of a Railroad Engineer* (Philadelphia: Railway World, 1896), 24; 16 March 1835, Managers' Minutes, Philadelphia and Reading Railroad, HML (hereafter P and R MM).

and the entire length to Pottsville open in January 1842.[162] English and American locomotives were used, and the double-tracked line was designed to accommodate seventy-five thousand tons of anthracite per day.[163] Richard Boyse Osborne, one of Robinson's assistants, and afterward his biographer, called the Philadelphia and Reading "the crowning achievement of [Moncure Robinson's] professional career."[164]

The construction of these later railroads demonstrates continued modification—in the direction of a unique American railroad technology—of the design Robinson employed for his earlier railroads. The track which he installed on the R, F and P and the Richmond and Petersburg was similar to that he had put down for his earlier railroads. The cross-ties and rails were of white oak, and the plate iron was nailed to the rails. The ties were laid directly on the earth, and there was no gravel ballasting.[165] As on the Petersburg this superstructure required substantial maintenance and repair.[166] The president of the R, F and P complained to Robinson in 1839 of "the almost incredible extent to which the roadway demanded repairs. The oak rails have not equalled our expectations, and the iron rail has proved too thin for the weight of our engines."[167]

Yet even with such difficulties this type of superstructure continued to be used on American railroads, particularly in the South, until the Civil War.[168] Moncure Robinson's use of it cannot be regarded as foolish if it had such a long life in the United States; others must have had similar reasons for using it. The plate rail without ballasting was the cheapest kind of superstructure that could have been laid, and minimally capitalized companies probably had no choice except to use it. If Robinson was willing to lay down a track which had a short life,[169] he may have been providing the only tech-

[162] Chevalier, *Histoire et description,* 2: 493.

[163] Brown, "Pioneer Locomotives," 56–57; Osborne, *Professional Biography,* 25.

[164] Osborne, *Professional Biography,* 24.

[165] Chevalier, *Histoire et description,* 2: 412, 417.

[166] Ibid., 2: 415.

[167] Jos. M. Sheppard, [April 1839], to Moncure Robinson [fragment], MRP.

[168] John Stover, *American Railroads* (Chicago: University of Chicago Press, 1961), 47; Eugene Alvarez, *Travel on Southern Antebellum Railroads, 1828–1860* (University: University of Alabama Press, 1974), 68–71.

[169] Besides the experience on the Petersburg, see Knight and Latrobe, *Report,* 4, 33, 37, 52.

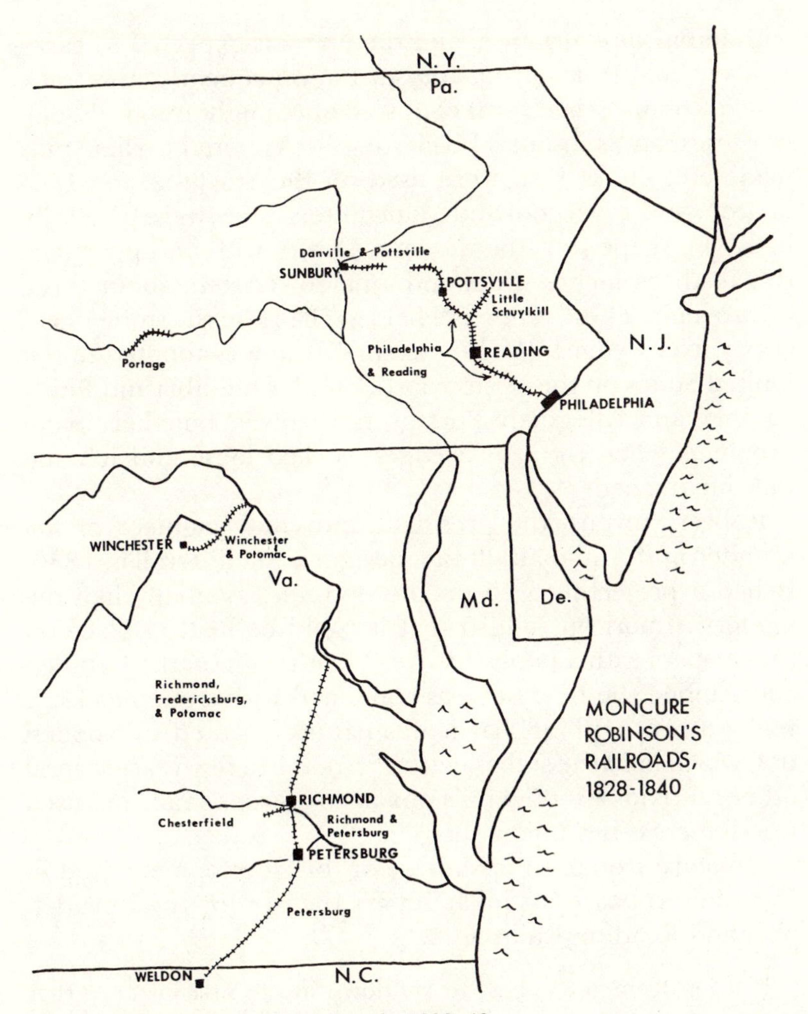

Fig. 11. Moncure Robinson's railroads, 1828–40.

nological alternative for the available capital.[170] The lower initial cost was apparently more important in such situations than the higher cost of repairs.

The Philadelphia and Reading Railroad had no problem

[170] On the R, F and P's lack of capital, see Chevalier, *Histoire et description,* 2: 412; Mordecai, *Brief History,* 9; Conway Robinson, 22 December 1839, to Moncure Robinson, MRP. In 1832 it was estimated that a double track plate iron railroad with wooden ties cost $11,751 per mile in the United States, and a double track line with edge rails and stone sleepers cost $18,550 per mile. See: Nicholas Wood, *A Practical Treatise on Railroads,* ed. George W. Smith (Philadelphia: Carey & Lea, 1832), 441–45.

with initial capitalization, and the line was expected to carry a heavy coal traffic from the start. Consequently, the track which Robinson laid for it consisted of wrought iron rails, oak cross-ties, and substantial ballasting.[171] As noted earlier, rails made entirely of iron were used on the Stockton and Darlington, and Liverpool and Manchester. They were generally in a "T" shape (not the modern T-rail), with an upper face two to three inches wide, and a narrower stem about three inches high. These rails were held in chairs on the stone sleepers by iron wedges.[172] This system of rail was adopted in the United States on the Boston and Lowell, Columbia and Philadelphia, and Allegheny Portage railroads,[173] but there were problems when the iron wedges worked loose and left the rails unattached.[174]

Robert Stevens, the president and chief engineer of the Camden and Amboy Railroad, designed a new T-rail in 1830. It had a projecting web on the bottom, essentially like the modern American rail, so that it could be held in place on the sleepers with a hooked spike.[175] Although on the Camden and Amboy the new rail was combined with stone blocks, it was soon recognized that it was naturally suited to wooden ties, which held the spikes better.[176] Within a few years several other railroads adopted rails similar to Stevens's rail, and used wooden cross-ties under them.[177]

Moncure Robinson was aware of this development, and in 1834 he proposed a similar superstructure for the Philadelphia and Reading Railroad.

> If it be deemed . . . best to lay down in the first instance that superstructure which, with reference to the large and ponderous trade to be accomodated by the improvement, which would seem to be the most advisable, I would recommend the following:

[171] Knight and Latrobe, *Report,* 20–21.

[172] Von Oeynhausen and von Dechen, *Railways in England,* 23–24, 44–45.

[173] Knight and Latrobe, *Report,* 11; "Columbia and Philadelphia Railroad. Report of William B. Mitchell, Superintendent," and "Allegheny Portage Railroad. Report of Sylvester Welch, Engineer," *Pennsylvania House Journal,* 1832–33, II, docs. 9, 15.

[174] E.g., W. H. Wilson, "Notes on the Philadelphia and Columbia Rail Road," *Journal of the Franklin Institute* 25 (May 1840): 336.

[175] J. Elfreth Watkins, "The Development of American Rail and Track," *Smithsonian Institution Annual Report, 1888–1889* (Washington: 1889): 666–68.

[176] Ibid., 670.

[177] Knight and Latrobe, *Report,* 10, 15–16, 23–24, 30–31.

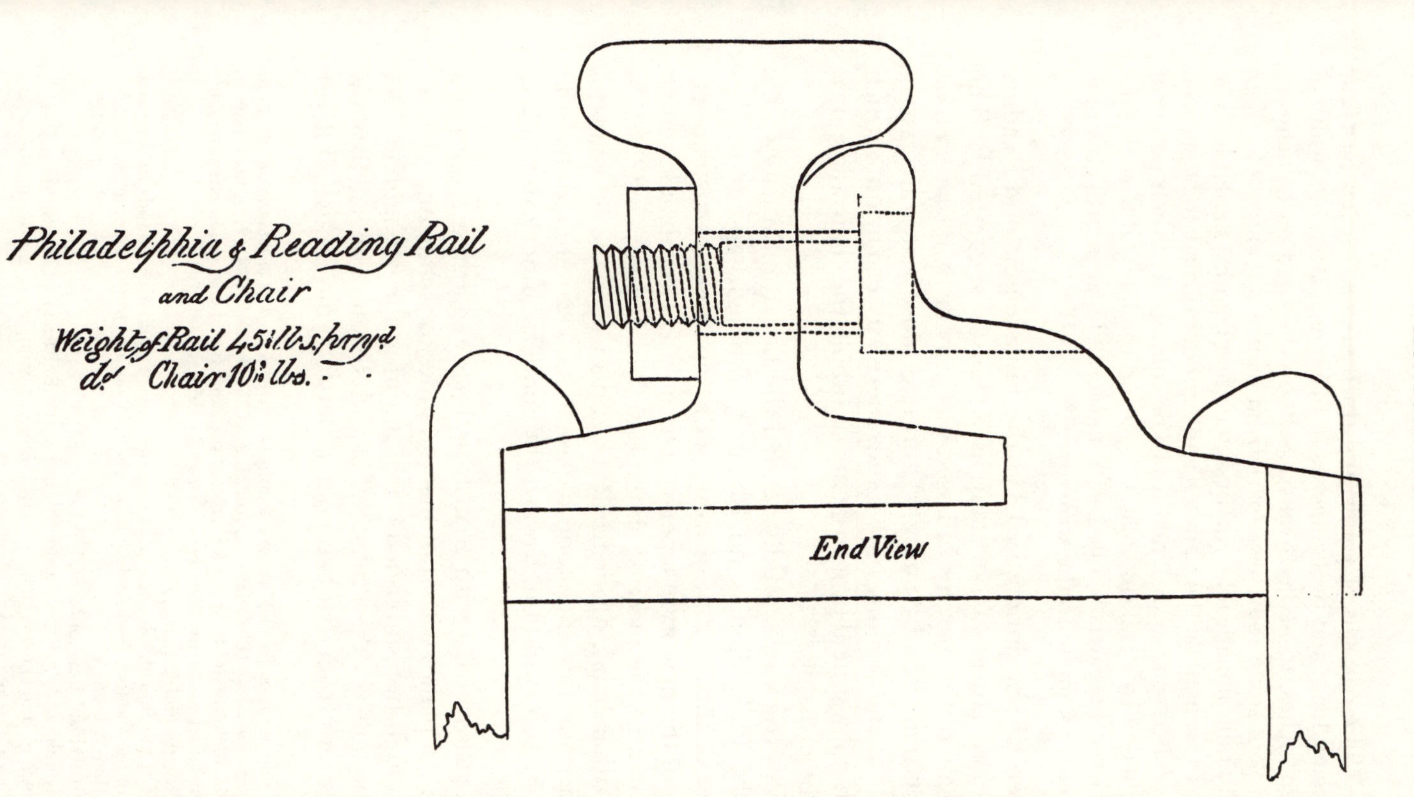

Fig. 12. This section of the Philadelphia and Reading rail was published in Jonathan Knight and Benjamin Latrobe Jr.'s *Report Upon the Plan of Construction of Several of the Principal Rail-Roads in the Northern and Middle States* (1838). Note that the rail was bolted to the chair, and that both were attached to the sleepers with hooked, or "billed," spikes. Courtesy of Hagley Museum and Library.

> White oak sills of young and thrifty timber, seven feet four inches long, and twelve inches in diameter flatted to a width of ten inches, to be laid transversely on the road on trenches of broken stone four feet apart from centre to centre. On these sills rails of malleable iron in sixteen feet lengths, similar in form to those used on the Camden and Amboy railroad, to be secured by billed spikes. . . . The rails to be connected at their points of junction by small side plates . . . , and the interval between the rails to be partially filled with moderately large broken stone, to give stability to the structure.[178]

When laid beginning in 1837 the Philadelphia and Reading superstructure was virtually as Robinson suggested.[179] The version of the Stevens rail used was heavier than that used on the Camden and Amboy,[180] and had more elaborate fastenings. At each joining of two rails there was a chair onto which each rail was held by a large nut-and-bolt; the chair was spiked to the cross-tie.[181]

This track successfully carried the heavy anthracite traffic of the Philadelphia and Reading, although (apparently because of the amount of iron used) Robinson's particular version of the new style superstructure "was soon abandoned on account of the increase in expense which it entailed."[182] On the other hand, Moncure Robinson's use of the Stevens rail with wood ties and stone ballast was a clear step away from the British tradition, toward the modern American rail and track.[183]

He also broke with British practice regarding grades and inclined planes with stationary engines (or self-acting planes). Although he established a maximum grade of eighteen feet to the mile for the Philadelphia and Reading, he had to allow variations from it at two points. Between the Schuylkill River

[178] The quote is from Moncure Robinson, *Report on the Continuation of the Little Schuylkill Rail Road* (Philadelphia: Joseph & William Kite, 1834), 7. In his report to the Philadelphia and Reading Railroad at the same time he recommended the same rail and track: Moncure Robinson, *Report* (n. 160 above), 8.

[179] Knight and Latrobe, *Report,* 20–22.

[180] On the Philadelphia and Reading the rail weighed 45⅛ pounds to the yard, and on the Camden and Amboy it weighed 41 pounds to the yard: ibid., 16, 20.

[181] Ibid., 21–22.

[182] Watkins, "Development," 672.

[183] Knight and Latrobe, *Report,* 20, found the Philadelphia and Reading "highly interesting" and appears to have drawn upon its example in considering the design of an ideal rail and track," 33, et seq.

at Fairmount and the Port Richmond terminal on the Delaware River there was a change in elevation of up to fifty feet to the mile, where he believed an extra locomotive would have to be on hand to assist trains, and near the Pottsville end there was a grade of twenty-six feet to the mile for three miles.[184] In his willingness to use steeper grades when absolutely necessary, instead of resorting to inclined planes, Robinson contributed to a spreading belief among American civil engineers that inclined planes were no longer useful. They recognized that the development of stronger and heavier locomotives permitted steeper grades (although sometimes more circuitous routes). Dispensing with inclined planes resulted in a shorter time of transportation and a reduced expense of operation.[185]

Robinson's later railroads also used bridges and tunnels more extensively than his early ones in order to reduce grades and eliminate inclined planes. The Philadelphia and Reading had three tunnels, and it and the R, F and P required a number of bridges to keep to low grades.[186] One may contrast the two large bridges at the termini of the Richmond and Petersburg, one of Robinson's later railroads, with Robinson's earlier inclined plane at the Roanoke River end of the Petersburg Railroad. His new solution to the problem of grades was to use bridges and tunnels where possible, and otherwise to depend upon locomotives to overcome steeper grades.

Certainly a key to these changes in railroad design was the development of a more powerful locomotive. What role Moncure Robinson played in that development is unclear. Daniel Calhoun has pointed out that almost from the beginning of American railroading civil engineers did not "commonly design and build locomotives."[187] Nonetheless, the claim has

[184] Osborne, *Professional Biography,* 33–35; Knight and Latrobe, *Report,* 20.

[185] Ferguson, *Early Engineering Reminiscences,* 152. The Baltimore and Ohio Railroad had an inclined plane at Parr's Ridge but stationary engines were never provided for it. In 1836 a new survey was made for a route to cross the ridge with locomotive engines "at a grade not exceeding seventy feet to the mile." See *Eleventh Annual Report . . . Baltimore and Ohio Rail Road Company* (Baltimore: Lucas & Deaver, 1837), 19. Also see "Report of J.P. Baily, relative to avoiding the Plane at Columbia," *Pennsylvania House Journal,* 1836–37, II, doc. 17; Benjamin Wright, 12 August 1834, to W. L. Marcy, Gratz Collection, HSP.

[186] See n. 160 above; Chevalier, *Histoire et description,* 22: 411–12.

[187] Calhoun, *American Civil Engineer,* 86.

been made that Robinson had some technical influence on the construction of the *Gowan and Marx,* a locomotive which represents an important step in American locomotive design. The *Gowan and Marx* (named after an English banking firm) was ordered by the Philadelphia and Reading Railroad in the summer of 1839 from the locomotive builders Eastwick and Harrison of Philadelphia. As completed it had two important innovations. The two pairs of driving wheels (which with the two pairs of leading wheels made it a "4-4-0") were positioned close together and well under the locomotive to put more of its weight on the driving wheels, a variation which was successfully tried on a locomotive for the Philadelphia, Germantown, and Norristown Railroad two years earlier. After that Joseph Harrison had perfected the "equalizing lever" which allowed the driving wheels to adjust to variations in the height of the rails, and thus maintain the stability of the locomotive much better.[188]

The *Gowan and Marx* was one of the earliest locomotives to combine these two innovations which were very important to American locomotives thereafter, and it was a remarkably successful machine. On one of its first trips over the railroad, made 20 February 1840, the *Gowan and Marx* "amazed the railroad world" by pulling a train of cars weighing 423 tons from Reading to Philadelphia at an average speed of 9.82 miles per hour.[189]

Richard Boyse Osborne, in his biography of Moncure Robinson, claimed the *Gowan and Marx* was "a locomotive of [Robinson's] design."[190] Michel Chevalier merely stated that the Philadelphia and Reading decided to order an eight-wheeled locomotive "following the advice of Mr. Robinson."[191] The minutes of the managers of the railroad reveal that in July 1839 the managers directed Moncure and Wirt Robinson to order a locomotive from Eastwick and Harrison. After the *Gowan and Marx*'s first trial (5 December 1839) the Robinsons reported that they had "given much attention to

[188] White, *American Locomotives,* 46–48, 287–89.

[189] George M. Hart, "History of the Locomotives of the Reading Company," *Railway and Locomotive Historical Society Bulletin* 67 (May 1946): 19.

[190] Osborne, *Professional Biography,* 36.

[191] Chevalier, *Histoire et description,* 2: 500.

the description of Engines most judicious for our Road," and wanted to work further with Eastwick and Harrison to design lighter and stronger locomotives.[192] Earlier the managers of the Philadelphia and Reading had routinely given responsibility for ordering locomotives to the Robinsons.[193] Until more evidence is uncovered it would be unwise to discount Moncure Robinson's role in the construction of the *Gowan and Marx* and, therefore, his impact on American locomotive design.

Yet most of the locomotives which were used on Moncure Robinson's railroads were built by British firms.[194] Locomotive historian John White believes that, from beginning to end, all the British locomotives shipped to the United States were unfit for conditions here. They were too heavy for the typically light track, had low-pressure engines which could not cope with the steeper grades on American railroads, and lacked the leading wheels, or "bogie truck," which helped to prevent rocking and derailment.[195] Yet the Philadelphia and Reading persisted in importing locomotives and in 1841 received the *Gem* "which is generally considered the last locomotive to have been imported from Britain."[196]

If Moncure Robinson was responsible for the ordering of locomotives for the lines on which he was chief engineer, and all evidence suggests that he was,[197] why did he continue in what appears a foolish course? In the early years his motivation may simply have been a strong attachment to British locomotives based on his belief in British engineering principles. But that does not explain why he continued to order British locomotives after he had moved away from British practice in rail and track. Although only twenty-five percent of the locomotives in the United States were of British manufacture in 1841,[198] at that date seventy-eight percent of the locomo-

[192] 17 July 1839, 11 December 1839, P and R MM.

[193] 15 April 1835, 16 September 1835, 26 April 1836, 5 September 1836, 17 July 1839, P and R MM.

[194] Only eight of thirty-two locomotives purchased up to 1841 by the Philadelphia and Reading, Richmond and Petersburg, and R, F and P were of American make: Brown, "Pioneer Locomotives," 56–57, 62–63.

[195] White, *American Locomotives*, 8.

[196] Ibid., 7.

[197] Note his assistant Wirt Robinson's trip to Britain in 1831–33 and Moncure Robinson's trip in 1836–37 (below) when they checked on the manufacture of locomotives.

[198] White, *American Locomotives*, 7.

tives which had been purchased by Robinson's seven locomotive-using lines were British.[199] Or, using another standard, about thirty-eight percent of all the British locomotives ever used in the United States were imported for Robinson's lines.[200]

It does not appear to have been engineering principle which made Robinson persist in an apparently foolish course. Robinson's later railroads probably imported British locomotives because a great deal of British capital was invested in them. Due to growing difficulties with borrowing money in the latter half of 1836,[201] both the Philadelphia and Reading and the R, F and P decided to seek capital in the London money market. Moncure Robinson was appointed the agent of both companies and sent to London to negotiate the sale of their stocks and bonds.[202] He left the United States late in November 1836,[203] and proceeded directly to London where he conferred with the banking house of Gowan and Marx. Through them he was able to market a million dollars worth of stock (twenty thousand shares) and a smaller amount of bonds for the Philadelphia and Reading Railroad.[204] At the same time he sold over three hundred thousand dollars worth of bonds of the R, F and P.[205]

Much of the proceeds of these sales and loans went toward the purchase of British rails and locomotives.[206] While in England Robinson personally checked into the manufacture of locomotives, apparently those intended for his Virginia lines.[207] He was also interested in inducing English mechanics and engineers to go to Richmond to work in the R, F and P shops. He attempted to find willing men at Liverpool and in London, but his letters indicate that he had little success.[208]

[199] Calculated from Brown, "Pioneer Locomotives," 29–30, 38–39, 49–52, 62–63, 76.

[200] Ibid.

[201] Taylor, *Transportation Revolution,* 342.

[202] Mordecai, *Brief History,* 9; 14 November 1836, P and R MM.

[203] In Charlotte Robinson, 6 December 1836, to Moncure Robinson, MRP, it is stated that he had been gone two weeks.

[204] Hare, *History of the Reading,* 6; 8 March 1837, P and R MM.

[205] Mordecai, *Brief History,* 17.

[206] Ibid.; 5 April 1837, P and R MM.

[207] Moncure Robinson, 5 February 1837, to Conway Robinson, MRP; Moncure Robinson, 2 March 1837, to Solomon White Roberts, The Charles Roberts Autograph Collection, The Library, Haverford College, Haverford, Penna.

[208] Ibid.

On this European visit Moncure Robinson spent several weeks in Paris. He renewed his acquaintance with Michel Chevalier, an eminent French civil engineer and public official whom he had met previously in the United States.[209] The Director-General of the Department of Bridges, Roads, and Mines invited him to a dinner with "a small group of railroad engineers."[210] While he was in Paris he was also contacted by the secretary of the Irish Railway Commission who had a list of questions for him.

> The Commissioners would be glad if you would enlighten them upon the economy of Railways, by which term, they would include, the comparative expense of first construction, under the various circumstances, of easy, or difficult country, double or single Lines, the employment of Locomotives, or Animal power, Costly purchases, or otherwise, of Land &c. The comparison of the subsequent cost of maintenance, and working, in the proportion to the Traffic.[211]

It is clear from this second visit of Moncure Robinson to Europe that a fundamental change had taken place in the relative position of his civil engineering knowledge to that in Europe. Ten years after completing his studies in Europe, which was recognized to be ahead of the United States in civil engineering skills in virtually all areas at the time, Robinson returned to be recognized as an authority on railroad construction. In the next few years government and private officials from Germany and Russia consulted with him concerning railroads, apparently recognizing that American economic and geographical conditions were more akin to theirs than were the British.[212]

Robinson's visit of 1836–37 also suggests a basic change in his personal career. No longer was he just a civil engineer; now he was involved in important administrative functions for his railroads. In handling the bond issues for the Philadelphia and Reading, and the R, F and P, he was entrusted

[209] Robinson, "Obituary" (n. 34 above), 36.

[210] E. Greaves, [n.d.], to [?], MRP. Also see: E. Greaves, 28 May 1837, to Moncure Robinson, MRP.

[211] Harry D. Jones, 27 April 1837, to Moncure Robinson, MRP.

[212] The Directors of the Kaiser Nordbahn, 1 May 1838, to Moncure Robinson, MRP; Osborne, *Professional Biography*, 37–38, 40.

with complex financial negotiations which to a large extent determined the success of those enterprises. At home he was spending more of his time in the office than in the field and relying upon his principal assistants to direct the field work, while he attended to problems of purchasing and organization.[213] Accompanying this change of function was Robinson's own heavy investment in his Pennsylvania and Virginia railroads to the extent that he was a large private stockholder.[214] The culmination of that involvement was his election to the presidency of the R, F and P in 1840, marking the virtual end of his civil engineering career.[215]

Moncure Robinson was one of the first civil engineers to make the change from engineering to administration, a path later followed by J. Edgar Thomson and John B. Jervis, among others.[216] As Eugene S. Ferguson has pointed out, the move to administration has been an indicator of an engineer's "success" ever since. In this case, with a substantial income derived from his engineering practice, Robinson became a major and influential investor in transportation routes along the eastern seaboard from Pennsylvania to South Carolina, and at his home in Philadelphia he frequently entertained business leaders.[217]

Although Moncure Robinson retired from civil engineering only twelve years after returning from France, his ideas continued to affect the development of American railroads through the men whom he had trained. From the time of his first work with the Pennsylvania Canal Commissioners in 1828 he had a "party" of assistants working with him,[218] and over

[213] This is apparent from the Wilson Miles Cary Fairfax diary, VHS.

[214] "List of Stockholders in the Richmond & Petersburg . . . 16 Feby. 1837," box 46, VaBPW; "A Small Stockholder," 24 April 1838, to Elihu Chauncey, MRP; 14 January 1840, Stockholders' Minutes, Philadelphia and Reading Railroad, HML.

[215] Mordecai, *Brief History,* 20, 82. Osborne, *Professional Biography,* 39–40, mentions two non-railroad projects on which Robinson was consulted up to 1847, when "he retired from the profession."

[216] "John B. Jervis," *DAB*; "John Edgar Thomson," *DAB.* Thomson's rise to power with the Pennsylvania Railroad is outlined in James A. Ward's "Power and Accountability in the Pennsylvania Railroad, 1846–1878," *Business History Review* 49 (Spring 1975): 37–59.

[217] Osborne, *Professional Biography,* 42–45; [Beverly Robinson], 8 April 1921, "To My Children and Grandchildren," box 9, series B, Group 10, Winterthur Manuscripts, HML. Ferguson's comment was made to the author.

[218] Moncure Robinson, 28 May 1828, to Joseph McIlvaine, box 1, Surveys & Correspondence, PaCC. A few months later Moncure referred to the "16 or 18

the years a large number of men served with his corps of assistants. A few, such as his cousin Wirt and a Frenchman ("Marcel"), were with Robinson for most of his professional life.[219] At first his assistants worked directly under him but after the spring of 1830 when he had concurrent responsibility for two or more railroads, he began to appoint "principal assistants" to whom he delegated authority while he was absent.[220] Through those close associates Robinson trained a much larger group of men than he could have done personally.

His assistants performed various tasks for him, but typically they carried out the more basic and laborious work, such as carrying surveying equipment or performing simple surveys, drawing maps and charts from survey data, calculating the expenses of constructing a railroad once Robinson had determined the route, and acting as "resident engineers" on location once construction was underway.[221] One of his former assistants later said of Robinson's style of training that:

> It has been repeatedly asserted by engineer officers in Mr. Robinson's employ, that his system, and discipline, so well carried out, supplied an education of the very best description for civil engineers. The requirement of strict obedience to general orders did not deprive them as officials of a proper manly independence, and they were left full opportunity for the exhibition of professional talent, in carrying out the same, each in his own way. Able officers in charge of the assistant engineers were willing teachers and advisors. . . .[222]

Of the outstanding engineers who were trained under Moncure Robinson the closest to him, and his longest associate,

hands as well as my assistants" who had made up his party: Moncure Robinson, 26 October 1828, to John Robinson, MRP.

[219] Early references to Marcel are in Moncure Robinson, 5 January 1828 [error for 1829], to Joseph McIlvaine, box 1, Surveys and Correspondence, PaCC; and Moncure Robinson, 18 August 1828, to John Robinson, MRP. The latest note I have seen states that he was "sinking a Victim to Intemperance": 10 March 1837, W. M. C. Fairfax diary, VHS.

[220] Ibid., 15 March 1835; "Officers, Engineers and servants of the Richmond and Petersburg Railroad Company . . . 1 Decem. 1837," box 46, VaBPW; Moncure Robinson, 2 October [1830], to Wirt Robinson, Acc. 1520, HML; Mordecai, *Brief History*, p. 83; *Hazard's Register of Pennsylvania* 16 (1835): 401.

[221] Entries for 15 March 1835 to 20 March 1838, Wilson Miles Cary Fairfax diary, VHS.

[222] Osborne, *Professional Biography*, 41.

was Wirt Robinson, who in 1839 succeeded him as chief engineer of the Philadelphia and Reading.[223] But there were others, such as Walter Gwynn who worked with Moncure Robinson on the Petersburg, then became chief engineer of a number of railroads in Virginia and North Carolina.[224] John H. Hopkins, an assistant on the R, F and P, became chief engineer of the Louisa Railroad, the oldest section of what became the Chesapeake and Ohio Railroad.[225] Richard Boyse Osborne, who began his career on the Philadelphia and Reading in the 1830s, later built a railroad in Ireland, then returned to build a number of American railroads.[226] William Hasell Wilson, who also was on the Reading under Robinson, had a long career with the Pennsylvania Railroad.[227] William S. Campbell, who was principal assistant on the Danville and Pottsville Railroad, became chief engineer on the Alabama, Florida, and Georgia Railroad.[228]

This pattern of training was common in the United States, as Daniel Calhoun has pointed out. Although prominent early American civil engineers, like Benjamin Henry Latrobe and Loammi Baldwin, Jr., had taught young men the profession through a personal apprenticeship, what had developed by Moncure Robinson's time was "the organized, hierarchical engineer corps, typically within a corporation or a state department of public works."[229] Through this system the technical knowledge which was brought from Britain by Moncure Robinson and other visiting American civil engineers was

[223] Above, p. 136; Hare, *History of the Reading,* 4, 21, 25, 32, 78. One of Moncure Robinson's conditions for accepting the position with the Philadelphia and Reading Railroad was that Wirt be hired at the same time: 30 March 1835, P and R MM.

[224] Moncure Robinson, 2 October [1830], to Wirt Robinson, Acc. 1520, HML; Moncure Robinson, 4 November 1830, to John Robinson, MRP; George W. Cullum, *Biographical Register . . . U.S. Military Academy,* 3 vols. (Boston: Houghton, Mifflin and Company, 1891), 1: 280–82.

[225] Moncure Robinson, 31 January 1835, to Board of Public Works of Virginia, (uncatalogued) Moncure Robinson box, VaBPW; Moncure Robinson, 5 February 1837, to Conway Robinson, MRP; Chevalier, *Histoire et description,* 2: 412; Charles W. Turner, "The Louisa Railroad, 1836–1850," *North Carolina Historical Review,* 24 (January 1949): 37, 39.

[226] "Richard Boyse Osborne, C.E.," *Engineering News* 42 (December 1899): 394.

[227] Wilson, *Reminiscences of a Railroad Engineer,* 44–55.

[228] *Hazard's Register of Pennsylvania* 16 (1835): 402; Chevalier, *Histoire et description,* 2: 433.

[229] Calhoun, *American Civil Engineering,* 47–48.

taught to a much larger group of men and diffused throughout the nation.[230]

There are a number of things about Robinson's transfer of British railroad technology to the United States which have significance for a general study of the transfer of technology. It is important to consider that he did not go to Europe in 1825 merely as an educated and inquisitive young man intent on steeping himself in the Old World culture. Rather he went as a skilled surveyor and civil engineer who was familiar with the largest civil engineering project in the United States up to that time—the Erie Canal—and as the successful engineer of a section of another major work—the James River canal. With such knowledge and skills he was well equipped to make an accurate assessment of the applicability of British railroad technology to American conditions, and, moreover, to know what in addition to his civil engineering knowledge he needed in order to build railroads in the United States.

Moncure Robinson's assessment of railroads as a better means of transportation than canals clearly dates from his initial visit to Britain in 1825 when he was first able to observe them. When he went to Europe he was concerned only with canals; he was not responding to any social or economic pressure for railroad engineers. This circumstance demonstrates the inadequacy of the frequent generalization that social or economic demand calls innovation into being.[231]

On the other hand, there was knowledge of, and interest in, railroads in the United States well before the technical skill for their construction was available. Printed materials and visitors to Europe brought the fascinating news of British railroad successes across the Atlantic. Entrepreneurs and capitalists became interested, and when the civil engineers who had visited Britain returned, they were quickly employed. Additionally, the common technological heritage of the two countries meant that there were in the United States men who could make the equipment which was not imported, and repair or (later) duplicate that which was. The American economy

[230] Stapleton, "The Origin and Diffusion of American Railroad Technology, 1825–1840."

[231] Eugene S. Ferguson, "Toward a Discipline of the History of Technology," *Technology and Culture* 15 (January 1974): 20.

and culture of the time provided the right soil for the seed of railroad technology to germinate.

After his successful transfer of British railroad technology to the United States due to his previous skills and the receptiveness of the American culture, Moncure Robinson at first closely adhered to British practice in constructing his railroads. He learned, however, to alter some elements of British practice to fit with American economic and geographic conditions, and took some major steps in the direction of a distinct American railroad technology, particularly in the development of railroad track.

Among the few men who brought British railroad technology to the United States Moncure Robinson ranks as one of the most successful and influential. Throughout his career he was ambitious and energetic, always wanting to be known as one of the best in his profession. He actively promoted railroad projects, and once he was appointed chief engineer, he attempted to build the kind of railroad which was appropriate to both the location and the financial capabilities of the line. In an era when many transportation projects failed, only one of Moncure Robinson's railroads, the Danville and Pottsville, was unsuccessful. Since he exhibited business acumen from the start, it is not surprising that he "retired" from civil engineering at an early age to devote himself to administrative pursuits. The railroad skills which he had transferred from Britain and later modified had, however, already spread widely in the United States.

V: David Thomas and the Anthracite Iron Revolution

Perhaps one of the most critical transfers of technology to the United States in the nineteenth century was the transfer of the process for smelting iron ore with anthracite coal beginning in 1840. Before that transfer the United States had only charcoal-fueled furnaces with cold, low-pressure blasts and (with few exceptions) bellows powered by waterwheels. Ironmaking technology was little changed from that of the first American furnace at Saugus, Massachusetts, in the seventeenth century.[1] While these charcoal furnaces generally produced a high-quality product which was well suited for agricultural implements and some machinery,[2] the growing market of the 1820s and 1830s was for cheap iron construction materials (e.g., rails for railroads, pipes for waterworks, columns and fronts for buildings, and steam engines). The American iron industry did not immediately respond to this change in demand and large amounts of iron were imported from Britain.[3]

Key to footnote abbreviations

DM = Board of Directors' Minutes, Lehigh Crane Iron Company, Accession 1198, Hagley Museum and Library, Greenville, Wilmington, Delaware.
JFI = *Journal of the Franklin Institute.*
PMHB = *Pennsylvania Magazine of History and Biography*
SM = Stockholders' Minutes, Lehigh Crane Iron Company, Accession 1198, Hagley Museum and Library, Greenville, Wilmington, Delaware.

---

[1] See Arthur Cecil Bining, *Pennsylvania Iron Manufacture in the Eighteenth Century* (Harrisburg: Pennsylvania Historical Commission, 1939), especially 71–83, for blast furnace technology before 1840.

[2] Louis C. Hunter, "Influence of the Market upon Technique in the Iron Industry in Western Pennsylvania up to 1860," *Journal of Business and Economic History* 1 (February 1929): 281. Although Hunter's argument pertains to a particular region, his description of the market can be applied to the entire United States.

[3] Peter Temin, *Iron and Steel in Nineteenth-Century America: An Economic Inquiry* (Cambridge, Mass.: M.I.T. Press, 1964), 21–22. The estimates of Robert W. Fogel in *Railroads and American Economic Growth* (Baltimore: Johns Hopkins Press, 1964), especially 194, show a decreasing but continued dominance of the American rail market by British suppliers from 1840 until the latter 1850s.

With the transfer of anthracite iron technology the technique and character of the American iron industry began to change, and in the four or five decades after 1840 the industry was revolutionized. By the end of the period coal-fueled furnaces produced the vast majority of American iron; charcoal furnaces were considerably different in technique; and the United States had become the world's leader in iron furnace design and operation.[4] David Thomas, a Welsh ironmaster who came to the United States in 1839 to build the first successful anthracite iron furnace, played a significant role in transferring the anthracite iron technology.

Thomas acquired his skill and knowledge in the British iron industry, which was the most technically advanced iron industry of the world in the early nineteenth century. Its position was due to three innovations. In 1709 at Coalbrookdale, Abraham Darby made the first successful attempt to manufacture iron with coked coal and thereafter there was a slow growth of the use of coke in Britain. No longer were all ironmasters dependent on charcoal from the shrinking forests. A second innovation was the use of steam power to blow the blast, first achieved at Brosely in 1776 by John Wilkinson. Steam power was not affected by seasonal variations of water supply, as water power was, and often provided more power than available water power. The third critical innovation occurred in 1828 when James Neilson patented the use of a heated blast, which reduced the amount of furnace fuel which merely heated the furnace air to a temperature suitable for successful combustion. Neilson's invention dramatically increased the productivity of British iron furnaces.[5]

Of these three innovations, the one which was the most radical was the use of coke. Darby's trial of coke fuel at Coalbrookdale was successful partly because of the serendipitous use of low-sulfur coal (sulfur being a damaging impurity in iron), and partly because coke iron was appropriate for the

[4] Temin, *Iron and Steel*, 96–98, 157–63; Richard H. Schallenberg and David A. Ault, "Raw Materials Supply and Technological Change in the American Charcoal Iron Industry," *Technology and Culture* 18 (July 1977): 436–66.

[5] Thomas Southcliffe Ashton, *Iron and Steel in the Industrial Revolution*, 2nd ed. (Manchester: Manchester University Press, 1951), 29–38, 69–70; W. K. V. Gale, *The British Iron and Steel Industry: A Technical History* (Newton Abbot: David & Charles, 1967), chaps. 2, 4, 9.

casting process invented by Darby. But it is unlikely that Darby would have attempted coke-firing at all had he not had experience with coke in the copper-smelting and beer-brewing industries. Even with that experience there must have been a protracted learning period after the first success at Coalbrookdale, which may be the best explanation for the long period before significant numbers of British furnaces turned to coke.[6]

In any case, the British experience with coal-fueled furnaces was unlikely to be transferable to the United States until its coal was sufficiently exploited and available in quantity at cheap prices. That happened by about 1825 when the Lehigh and Schuylkill navigations tapped the anthracite coal resources of northeastern Pennsylvania.[7] The lack of cheap coal also tended to limit the application of another of the British innovations, the introduction of steam engines for blowing the blast, because fuel for the engines was a major expense. As late as the 1830s there were few furnaces in the United States with steam engines.[8] On the other hand, there was no resource obstacle to speedy experimentation with the third innovation, the hot blast, but when it was tried at a charcoal furnace in New Jersey in 1835, it was not a great success.[9] It did not come into common use until after 1840.[10]

[6] Hyde has assembled figures showing that up to the 1740s the Coalbrookdale furnace and forge had higher production costs than comparable charcoal-fueled works. I would attribute that in some degree to the slow acquisition of knowledge about how to operate a coke-fueled furnace efficiently. Charles K. Hyde, *Technological Change in the British Iron Industry, 1700–1870* (Princeton: Princeton University Press, 1977), 32–41.

On the non-verbal nature of coal-fuel technology and the learning of its mysteries, see the seminal article by John Harris, "Skills, Coal, and British Industry in the Eighteenth Century," *History* 61 (June 1976): 167–82.

[7] Alfred D. Chandler, Jr., "Anthracite Coal and the Beginnings of the Industrial Revolution in the United States," *Business History Review* 46 (Summer 1972): 163. A readable account of the anthracite district before 1825 is provided by H. Benjamin Powell, *Philadelphia's First Fuel Crisis: Jacob Cist and the Developing Market for Pennsylvania Anthracite* (University Park: The Pennsylvania State University Press, 1978).

[8] Two usually reliable sources suggest that few steam engines were in use at blast furnaces before 1840. James M. Swank, *History of the Manufacture of Iron in All Ages* (Philadelphia: published by the author, 1884), 147, 160, 172, mentions three. The "steam engine census" of 1838 lists eight furnaces with steam engines: "Steam Engines," 25th Cong., 3rd Sess., HR doc. 21 (1838) [serial set no. 345], 133, 156, 183, 191–192, 410.

[9] Swank, *History,* 326; W. David Lewis, "The Early History of the Lackawanna Iron and Coal Company," *PMHB* 96 (October 1972): 427, 429.

[10] Swank, *History,* 326.

While ample supplies of cheap coal were a precondition for the transfer of British coal-fuel technology to America, experienced British iron workers actually transferred the technology. There was the added dimension, at first, that only anthracite coal, not bituminous, was abundantly available for American furnaces, and anthracite was impossible to coke in preparation for the furnace as bituminous coal was in Britain. Since American bituminous coal was not mined in quantity until well after the anthracite beds of northeastern Pennsylvania were opened, the early attempts to smelt with mineral fuel focused on anthracite.

W. Ross Yates has reviewed the American experiments and their indifferent results.[11] As early as 1826 the Lehigh Coal and Navigation Company (operators of the Lehigh Canal) built a furnace at Mauch Chunk, Pennsylvania, in which anthracite was used as a fuel without success. The next year an ironmaster of Lebanon, Pennsylvania, published a notice of experiments with anthracite which discussed problems with blast pressure and furnace size.[12] Then, in 1831 after Neilson's hot blast had been invented, Frederick W. Geissenhainer performed some successful experiments in a small hot-blast furnace. On that basis he received a patent which covered the use of both the cold and hot blast for the manufacture of iron with anthracite. He attempted commercial production in 1836 when he built a furnace in Schuylkill County, Pennsylvania, but after two months his hot-blast machinery broke down and the work stopped. Geissenhainer died in 1838 before he could carry out further experiments.

Late in the 1830s there were two further trials. In 1838 the firm of Baughman, Guiteau and Co. built a small furnace near Mauch Chunk and produced iron with anthracite and a hot blast, but output was small and the company went bankrupt.[13] In the next year the Pioneer furnace at Pottsville was

[11] W. Ross Yates, "Discovery of the Process for Making Anthracite Iron," *PMHB* 98 (April 1974): 206–23. The following discussion of anthracite blast furnaces before mid-1840 is based on pp. 208, 210–11, 217, 219–20 in Yate's article.

[12] Joshua Malin, "Description of a Furnace for Smelting Iron, by means of Anthracite," *The Franklin Journal and American Mechanics Magazine* 4 (October 1827): 217–19.

[13] The Easton (Penna.) *Democrat & Argus* of 17 September 1840 and succeeding weeks carried a notice for a sheriff's sale which describes this furnace thoroughly.

put into blast and remained in production for ninety days, but it had many difficulties with its hot-blast machinery and did not operate for some time afterward. It is interesting that the Pioneer furnace was superintended by a Welshman, Benjamin Perry, who had experience with ironmaking in his native country.[14] Perry was involved with the blowing-in of two more anthracite furnaces in 1840, and there were two others in production before the middle of that year in northeastern Pennsylvania.

Yet none of these attempts to smelt iron with anthracite was a complete technical success. All shut down within a year because of low production, poor-quality iron, and problems with the blast machinery.[15] The correct mixture of furnace size, blast pressure, temperature of the blast, and other elements had not been achieved. It was on the other side of the Atlantic that the proper method was first found.

There is in South Wales near Swansea a bed of coal with strong anthracitic characteristics; that is, it is coal with a very high carbon content, a low level of impurities (especially sulfur), and very little matter which can be passed off as a gas when heated. In the coking process by which coal was prepared for the furnace in Britain in the eighteenth and early nineteenth centuries, coal was charred like charcoal until the gaseous matter had escaped.[16] The resulting coke was porous, which allowed better air circulation and combustion in the furnace. The anthracite coal in South Wales, however, just like that in northeastern Pennsylvania, had such a small quantity of potential gas that it did not coke properly, and was considered useless as furnace fuel by the neighboring iron works.

The Yniscedwyn Ironworks were located in the middle of the anthracite field about thirteen miles from Swansea. Be-

[14] In addition to Yates, "Discovery," 219, see *The Miners' Journal* (Pottsville, Penna.), 26 October 1839, and Ele Bowen, ed., *The Coal Regions of Pennsylvania* (Pottsville, Penna.: Carvalho and Co., 1848), 32.

[15] Yates, "Discovery," 220–21; Swank, *History,* 273. For example, the Pioneer furnace, which received a prize for its ninety-days' production, had major problems with its hot blast: *The Miners' Journal,* 18 July 1840, 1 August 1840, 12 September 1840; S. W. Roberts, 26 April 1841, to Charles Roberts, Roberts Autograph Collection, Haverford College Library, Haverford, Penna.

[16] Gale, *British Iron and Steel Industry,* 30–31; Ashton, *Iron and Steel,* 31; Temin, *Iron and Steel,* 59, 89, 201

cause of its location coke had to be brought to it from more than ten miles away.[17] The works had been in existence producing good iron since sometime in the eighteenth century,[18] and in 1823 were acquired by George Crane, who had retired from the hardware business in Birmingham.[19] He retained the superintendent of the works, David Thomas, who had come to Yniscedwyn in 1817 after five years' experience in iron manufacture at the nearby Neath Abbey Ironworks,[20] one of the leading works in Britain.[21]

Although there must have been many other skilled iron workers at Yniscedwyn, David Thomas was the superintendent. He was the only son of a farmer of Tyllwyd in Glamorganshire, Wales, and had been given the best schooling available in his neighborhood. In 1812, at age seventeen, he began work for the Neath Abbey Ironworks near Swansea, and according to his biographer, "here young David spent five years, acquiring his technical training in the machine-shops and foundry, while devoting his leisure hours to the study of the working of the blast furnaces."[22] In 1817 he was in Cornwall for several months erecting a pumping engine. In that same year, perhaps because of Neath Abbey's change of ownership, he moved to the Yniscedwyn Ironworks to become its general superintendent.[23]

[17] George Crane, "On the Smelting of Iron with Anthracite Coal," *JFI* new ser. 21 (1838): 127–28; David Thomas, 23 February 1872, to F. H. Lynn, in *Guide-Book of the Lehigh Valley Railroad* (Philadelphia: J. B. Lippincott and Co., 1873), 154–56 (hereafter cited as Thomas to Lynn, *Guide-Book*).

[18] W. E. Minchinton, "The Place of Brecknock in the Industrialization of South Wales," *Brycheiniog* 7 (1961): 23.

[19] Solomon W. Roberts, "Obituary of the late George Crane, Esq., the founder of the Anthracite Iron Manufacture," *JFI* new ser. 11 (1846): 214. In Britain the "hardware business" included founding and machine manufacture.

[20] Edward Roberts, "David Thomas: The Father of the Anthracite Iron Trade," *Red Dragon* (October 1883), issued separately (Westfield, Neath., [1883]), 4.

[21] Lawrence Ince, "The Neath Abbey Ironworks," *Industrial Archeology* 11 (Spring 1977): 21–37.

[22] Roberts, "David Thomas," 3–4. According to Ince, "Neath Abbey Ironworks," 25, 29, Neath Abbey was a notable producer of steam engines during Thomas's training there.

[23] Ince, "Neath Abbey Ironworks," 25; Roberts, "David Thomas," 4. Roberts appears to be reliable as a biographer. He interviewed one of David Thomas's sisters, and Samuel Thomas, David Thomas's son, stated that "Mr. Roberts's narrative is authentic, by reason of his free access to the records of the Yniscedwyn anthracite iron-works during his long connection with them." Samuel Thomas, "Reminiscences of the Early Anthracite-Iron Industry," *Transactions of the American Institute of Mining Engineers* 29 (1899): 902.

Fig. 13. David Thomas, ca. 1870. Courtesy of Hagley Museum and Library.

The combination of Crane and Thomas was a dynamic one, and they began in 1826 to experiment with anthracite coal.[24] Thomas later stated that a small experimental furnace was built in 1825 for the trial in the next year, and it was enlarged

[24] Crane, "On the Smelting of Iron," 127; Thomas to Lynn, *Guide-Book.*

and charged with anthracite again in 1832. Neither attempt was a success.[25] Sometime later, probably in 1836, the idea occurred to Thomas and Crane that the Neilson hot blast which was coming into wide use in Scotland might solve their problem.[26] David Thomas went to Scotland to learn about the new technique, and returned with a license from Neilson to use it, as well as a mechanic to build a hot-blast oven.[27] There was some preliminary experimentation with the combination of anthracite and the hot blast, but it was early in February 1837 before Crane and Thomas successfully produced iron with anthracite at Yniscedwyn.[28] Something of the excitement of the moment was communicated by Crane in a letter to one of the proprietors of a neighboring ironworks in a letter dated 14 April 1837.[29]

> Have you heard that I have now successfully brought the Stone Coal [i.e., anthracite] question to a termination. The experiment cost me the labor of some months, and has been attended with a serious expense, but I am now making the Ton of Pigs with 31 to 33 Cwt. of Raw Stone Coal, the fuel 10 to 11 Cwt. for the Heat [sic] Air Stoves to be added, and we hope to do better than this. The Iron *very much* stronger than *cold blast* Coke Iron. I am told that Mr. Crawshay and yourselves are forming New Works in the Forest of Dean, if you were to come down to see what I have now been doing for the last *8* or *9* weeks with Stone Coal only, if it is not too late, I think that you would recommend the Dowlais Company to apply for a license to me, and look out for some eligible spot in some part of the Stone Coal District . . . .

Thereafter anthracite coal was the only fuel used at the Yniscedwyn furnaces until they ceased production in 1877.[30]

Yniscedwyn was clearly successful in smelting iron with anthracite because of the application of the hot blast. Crane and

[25] Thomas to Lynn, *Guide-Book.*

[26] Crane told the story of how the idea occurred in his article, "On the Smelting of Iron," 129; Thomas's story in Roberts, "David Thomas," 6.

[27] Roberts, "David Thomas," 6.

[28] Crane, "On the Smelting of Iron," 127; Thomas to Lynn, *Guide-Book.*

[29] George Crane, 14 April 1837, to Thomas or John Evans, in Madeleine Elsas, ed., *Iron in the Making* (Frome and London: Glamorgan County Council, 1960), 202–203. Crane's reference to a "license" concerns the patent on the process which he had obtained. See "Specification of a Patent for Smelting Iron with Anthracite Coal," *JFI* new ser. 21 (1838): 405–406; Roberts, "Obituary," 216.

[30] Minchinton, "The Place of Brecknock," 23–24.

Thomas were among the innovators in its use in their region, since a recent estimate is that "as late as 1839, only one-sixth of the Welsh furnaces were using the hot blast."[31] The blast at Yniscedwyn was heated outside the furnace, as Crane's letter indicates.[32]

Another element of the success was the size of the furnace at Yniscedwyn. It was forty-one feet high, ten and a half feet wide at the widest part (the "bosh"), and three feet six inches square at the bottom, or "hearth," where the furnace was tapped.[33] This was substantially larger than contemporary furnaces in the United States, including those used in attempts to smelt iron with anthracite,[34] and as a result it probably had a better draft to aid in combustion. These dimensions also permitted greater charges and larger production.

The most important contribution to success at Yniscedwyn, however, was the British technical experience with coal in blast furnaces which was manifest in David Thomas. John Harris has pointed out the deep familiarity with coal-fuel techniques which the British had gained since the sixteenth century.[35] Quite a number of industries, including glass manufacture and copper smelting, had converted to coal before the nineteenth century through continued trial and adaptation. Smelting iron with coke rather than charcoal was almost universal in Britain by the time of the Yniscedwyn experiments, while in the United States only a few abortive attempts had been made.[36] Even the hot blast, although a recent invention, had already been modified by improving the stoves for heating the blast and by providing water-cooled tuyères, the nozzles

[31] Hyde, *Technological Change and the British Iron Industry,* 155.

[32] See n. 29 above, and Crane, "On the Smelting of Iron," 128.

[33] Crane, "On the Smelting of Iron," 127. Other sources give the height as forty-five feet and the width at the bosh as eleven feet: Roberts, "David Thomas," 6; Thomas, "Reminiscences," 904; Thomas to Lynn, *Guide-Book.*

[34] William Firmstone, "Sketch of Early Anthracite Furnaces," *Transactions of the American Institute of Mining Engineers* 3 (1874–75): 153–55; Walter R. Johnson, *Notes on the Use of Anthracite in the Manufacture of Iron* (Boston: Little and Brown, 1841), 28–29.

[35] Harris, "Skills, Coal and British Industry," 167–82; John R. Harris, "Saint-Gobain and Ravenhead," in *Great Britain and Her World, 1760–1914: Essays in Honour of W. O. Henderson,* ed. Barrie M. Ratcliffe (Manchester: Manchester University Press, 1975), 27–70; John R. Harris, *Industry and Technology in the Eighteenth Century: Britain and France* (Birmingham: University of Birmingham, 1972).

[36] Swank, *History,* 278–79.

which admitted the blast into the furnace.[37] When David Thomas went to Scotland to learn about the hot blast it was possible to bring back an "experienced mechanic" to construct the apparatus at Yniscedwyn.[38]

There has been some disagreement as to whether George Crane or David Thomas deserves the credit for the success at Yniscedwyn. In later years Crane's eulogizer emphasized his role, while Thomas's son defended his.[39] It appears that Crane supported and actively prosecuted the experiments since he was the proprietor of the works, but there is no evidence that Crane had the technical skill to carry out or assess the experiments made. Clearly that was the contribution of David Thomas, whose training at Neath Abbey, and subsequent experience at Yniscedwyn made him an expert in the British technique of iron smelting with coke.[40] In the context of the transfer of technology, Thomas was incomparably better prepared to experiment with anthracite than any American of the time.

News of the smelting of iron with anthracite in South Wales did not take long to cross the Atlantic. It was transmitted by Solomon White Roberts, an American civil engineer who had been in Wales since sometime in 1836 as an inspector of iron manufactured for American railroads.[41] Apparently he was in the employ of A. and G. Ralston of Philadelphia, who were major importers of British rails.[42] In May 1837, about three months after the first successful anthracite blast, Roberts visited Yniscedwyn and then notified his uncle Josiah White of his observations.[43]

[37] Hyde, *Technological Change and the British Iron Industry,* 154.

[38] Roberts, "David Thomas," 6.

[39] Roberts, "Obituary," 216; Thomas, "Reminiscences," 926.

[40] The Neath Abbey Ironworks used coke. Ince, "Neath Abbey Ironworks," 21, 23; Roberts, "David Thomas," 4.

[41] Solomon W. Roberts, *Memoir of Josiah White* (Easton, Penna.: Bixler & Corwin, 1860), 6; Solomon W. Roberts, "Reminiscences of the First Railroad Over the Allegheny Mountain," *PMHB* 2 (1878): 388; Sylvester Welch, 13 December 1836, to Charles Roberts, Roberts Autograph Collection; W. S. Campbell, 29 March 1836, to Moncure Robinson (extract), box 43, Acc. 1520, Hagley Museum and Library, Wilmington, Delaware.

[42] Moncure Robinson, 5 February 1837, to Conway Robinson, Moncure Robinson Papers, Earl Gregg Swem Library, The College of William and Mary in Virginia, Williamsburg, Va.

[43] Roberts, *Memoir,* 6; Solomon W. Roberts, "The Early History of the Lehigh Coal and Navigation Company," *The Railway World* 1 (1875): 299.

White was a manager of the Lehigh Coal and Navigation Company (LC and N), which operated the Lehigh Canal and in addition, owned and mined large tracts of anthracite coal lands in northeastern Pennsylvania. The LC and N had been actively encouraging experimentation with anthracite coal along the canal route. In 1835 the managers agreed to supply an unnamed local company with a thousand tons of coal "to be used exclusively in experiments in smelting iron ore within two years from the 1st of August 1835."[44] They also stipulated that if the company produced iron with anthracite for three months, it would also receive lower prices on the Navigation Company's coal and be able to ship certain amounts of coal or iron toll-free on the canal.[45] Shortly thereafter the LC and N offered similar benefits to any company attempting to smelt iron with anthracite, adding an offer of free water power for the experiment from the Lehigh Canal or one of the canal's dams. On proof of success, the LC and N would grant a deed to any water power between lock thirteen (near Palmerton) and dam eight (the last dam, near Easton), a prime industrial area. To fulfill the requirements for these bonuses the experimenter had to begin by 15 August 1835, produce iron for three months, and complete the experiment within two years.[46] In 1836 the managers of the LC and N repeated the offer, but the few companies which applied to take advantage of the terms never got underway.[47]

The state of Pennsylvania also provided some encouragement to experimenters by passing a special act in 1836 allowing companies using coke or coal in the smelting of iron to incorporate themselves, an advantage not generally enjoyed by industrial concerns at this time.[48] Yet even with this legislation, and with the publication of the news of Crane's success,[49] there were in the next months no successful attempts to make iron with anthracite.

[44] Anthony Joseph Brzyski, "The Lehigh Canal and Its Effect on the Economic Development of the Region through Which It Passed" (Ph.D. dissertation, New York University, 1957), 469.

[45] Ibid.

[46] Ibid., 469–70.

[47] Ibid., 471.

[48] "Articles of Association," 23 April 1839, SM; Louis Hartz, *Economic Policy and Democratic Thought: Pennsylvania, 1776–1860* (Chicago: Quadrangle Books, 1968), 39–40.

[49] E.g., "Stone Coal Iron," "Iron from Anthracite," *JFI* new ser. 20 (1837): 185;

The interest of the LC and N was still great, however, and in October of 1838 Nathan Trotter, one of the managers, reported that he had "been this afternoon engaged with the Board of Managers of the Lehigh Coal & Navigation Company and that subject of Smelting Iron with Anthracite Coal was the subject of much conversation and is one in which the interest of the Company is much involved."[50] What apparently stimulated this discussion was the impending journey to England of another of the managers, Erskine Hazard, to sell a bond issue of the company.[51] He was delegated by an informal group of investors, most of whom were also managers of the LC and N, to visit Yniscedwyn and assess the potential of the new process in the United States.[52]

At the same time the LC and N made a new offer to experimenters. It would grant "all the water power of any one of the dams between Allentown and Parryville" (within the area previously offered) to any company which capitalized itself at $50,000 or more and expended at least $15,000 on experiments within two years from 1 July 1839.[53] According to one historian these terms were intended to eliminate competition with the new company by smaller businesses.[54] That seems likely, since at the board meeting in which Hazard was delegated to go to Europe the organizers of the informal company concluded an acceptance of the Lehigh Company's proposal.[55]

Hazard made his trip in November of 1838. In the next month he was at Yniscedwyn, where he was soon convinced that the anthracite iron process could be transferred to Pennsylvania. In discussing the matter with him, George Crane recommended that Hazard employ David Thomas as ironmaster. Although Thomas was at first reluctant to emigrate, on 31 December 1838 he signed an agreement with Hazard to build a furnace on the Lehigh River for making iron with

Report of the Board of Managers of the Lehigh Coal and Navigation Company to the Stockholders (Philadelphia: James Kay, Jr. and Brother, 1838), 19.

[50] Nathan Trotter & Co., 3 October 1838, to Jevons Sons and Co., Foreign Letters Sent, Nathan Trotter Collection, Baker Library, Harvard University, Cambridge, Mass.

[51] Brzyski, "Lehigh Canal," 304–305, 477.

[52] Ibid., 477; 9 November 1840, SM.

[53] Brzyski, "Lehigh Canal," 476–77.

[54] Ibid., 477.

[55] Ibid.

anthracite. Hazard and Thomas then ordered machinery for one blast furnace to be shipped to the United States.[56] David Thomas's son later recalled that "the blowing-machinery was constructed at the Soho Works, England, and the hot-blasts at Yniscedwyn from the same patterns as used there . . . while the firebrick came from the Stourbridge works, England."[57]

Hazard returned to the United States in April and the corporation was formally organized as the "Lehigh Crane Iron Company" (LCIC) under the Pennsylvania act of 1836. Named in honor of George Crane (who had no investment in the company), the company was capitalized at $100,000. Its two thousand shares were subscribed by eight persons, including Josiah White and Erskine Hazard who each took four hundred shares. The LCIC was chartered by the state of Pennsylvania on 20 May 1839.[58] For its time it was a major enterprise and involved the risk of a large amount of capital, although due to the privilege of incorporation (rare at the time) the shareholders had liability limited to their investment.[59]

David Thomas, his wife, and three sons arrived at New York in May and remained there for a month because he was ill with a fever. In July he went to the site of the Lehigh Canal that the directors of the company had chosen. (It later became the town of Catasauqua.) After laying out the works Thomas directed the construction of the furnace, the hot blasts and the water-powered blowing machinery. The furnace was probably larger than any other in America at the time,[60] being forty-five feet high, twelve feet at the bosh, and four and a half feet across the hearth.[61] The hot blasts were, as at Yniscedwyn, separate from the furnace. A breast waterwheel twelve feet in diameter drove the blast machinery.[62]

[56] 9 November 1840, SM.

[57] Thomas, "Reminiscences," 927. The Lehigh Crane Iron Company later paid the Yniscedwyn Iron Company £600: 11 November 1839, DM.

[58] Brzyski, "Lehigh Canal," 305; "Articles of Association," 23 April 1839, SM; 23 May 1839, DM.

[59] One economic historian concludes that the shift from charcoal to anthracite fuel was accomplished mostly through major new investments which greatly increased the scale of the typical iron industry plants. William D. Walsh, *The Diffusion of Technological Change in the Pennsylvania Pig Iron Industry, 1850–1870* (New York: Arno Press, 1975), 160, 171–73.

[60] Firmstone, "Sketch," 155. The location was named "Catasauqua" in 1846: 2 March 1846, DM.

[61] Thomas, "Reminiscences," 908–909.

[62] Ibid., 908, 910–911.

There were, however, some difficulties in constructing what was called a "works . . . upon the Welsh plan" in the United States.[63] Although the cylinders (five feet in diameter) for the blast engine had been ordered in Britain, they failed to arrive with the other equipment, because the hatchway of the ship to which they were consigned was not large enough to receive them.[64] The LCIC tried to order new ones, but "at this time there was not a boring-mill in the United States large enough to bore a cylinder of sixty inches diameter."[65] Finally the new firm of Merrick and Towne of Philadelphia took the order and delivered satisfactory cylinders in May 1840.[66] Another problem encountered was locating a foundry in which to cast the large iron pieces needed for the waterwheel, but they were finally made nearby at Allentown.[67]

Once all the equipment was in place at the furnace "some partial trials [were] made."[68] These may have tested the ores available in the area and ascertained that all the equipment was working properly.[69] Nathan Trotter reported late in June that "we started our Anthracite Furnace and all went well enough until one of the tweyers [sic] gave out and occasioned some delay. There is however no doubt of its success."[70]

Finally, the furnace was filled on 2 July, put in blast the next day, and produced nearly twenty tons of iron in its first week. Over the first seventeen weeks it averaged over forty-one tons per week, a large output compared with contem-

[63] *A History of the Lehigh Coal and Navigation Company* (Philadelphia: William S. Young, 1840), 58; *Report of the Lehigh Coal and Navigation Company* (Philadelphia: William S. Young, 1840), 28.

[64] 9 November 1840, SM; Thomas, "Reminiscences," 910.

[65] Thomas, "Reminiscences," 912.

[66] 9 November 1840, SM; Thomas, "Reminiscences," 913. Although there is no direct evidence to corroborate the story of the enlargement of the boring mill, it is not inconsistent with published descriptions of the Merrick and Towne "Southwark Foundry" established in 1839. William Hamilton, "Report on the Boring Mill, Constructed by Messrs. Merrick and Towne," *American Railroad Journal* 11 (1840): 186–88; J. Leander Bishop, *A History of American Manufactures*, 2 vols. (Philadelphia: Edward Young and Co., 1864) 2: 547–48; Bruce Sinclair, *Philadelphia's Philosopher Mechanics: A History of the Franklin Institute, 1824–1865* (Baltimore: Johns Hopkins University Press, 1974), 290–91.

[67] Thomas, "Reminiscences," 914.

[68] 9 November 1840, SM.

[69] Various ores from Pennsylvania and New Jersey were used at first: 9 November 1840, SM.

[70] Nathan Trotter, 30 June 1840, to Jevons Sons and Co., Foreign Letters Sent, Nathan Trotter Collection.

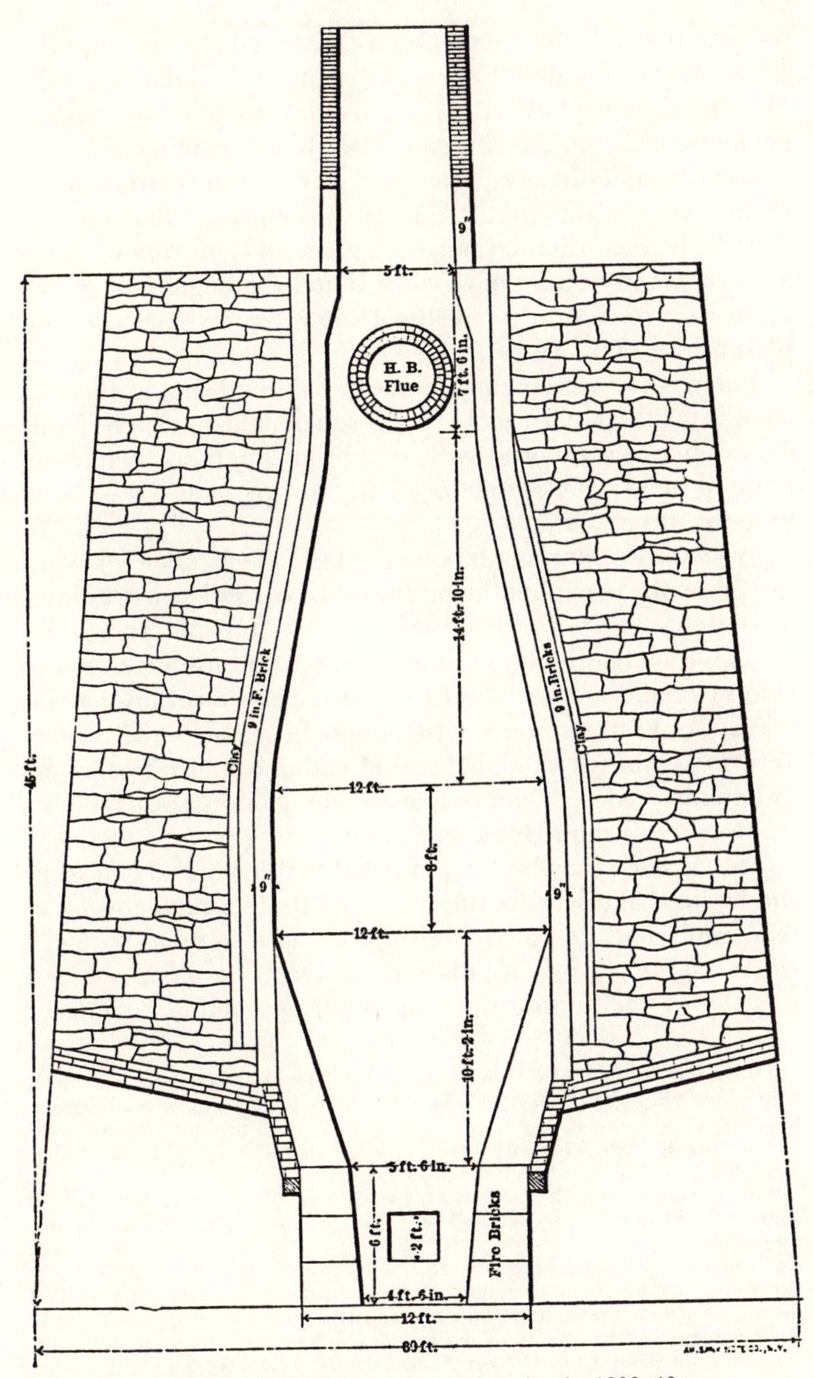

Fig. 14. Furnace No. 1, built at Catasauqua, Pennsylvania, 1839–40.

porary American furnaces, and it continued in blast until the Lehigh River flooded the works in January and damaged the blast machinery.[71] By May, when Thomas put the furnace back in operation, the company was already making plans for a larger second furnace,[72] having received enthusiastic reports from various places on the quality of the iron.[73] The work of David Thomas and the Lehigh Crane Iron Company was thus a success, and writers have since dated the beginning of the anthracite iron industry in the United States from the first blast at the LCIC works.[74]

Thomas's achievement in building and blowing-in the furnace at Catasauqua can be put in proper perspective if his experience is compared with two other attempts to blow-in mineral fuel furnaces: one prior to blowing-in at Catasauqua and one after.

In 1836 two Marylanders formed the Georges Creek Mining Company to exploit the bituminous coal and iron ore along a tributary of the Potomac River in western Maryland. Although they had no experience in the iron business they decided to build a coal-fueled iron furnace at a promising site known as Lonaconing. Construction began in 1837, and a Welsh ironmaster named David Hopkins was hired to superintend the furnace. Over the next several months other Welsh ironworkers joined Hopkins at Lonaconing.

The furnace they built was probably the most advanced in the United States at the time. It was fifty feet high and fourteen and a half feet at the boshes; it had a sixty-horsepower steam engine and two hot-blast ovens. Operations commenced on 16 May 1839 using raw (uncoked) bituminous coal (later

[71] 9 November 1840, 14 November 1841, SM; Thomas, "Reminiscences," 915; Solomon White Roberts, 26 April 1841, to Charles Roberts, Roberts Autograph Collection.

[72] 25 August 1840, 1 January 1841, 20 September 1841, DM; 14 November 1841, SM.

[73] 9 November 1840, SM; Nathan Trotter, 30 July 1840, 19 October 1840, 19 December 1840, to Jevons Sons and Co., Foreign Letters Sent, Nathan Trotter Collection.

[74] Temin, *Iron and Steel*, 61; Swank, *History*, 273; Yates, "Discovery," 220–21. An exhaustive survey of early attempts to smelt with anthracite has recently reached the same conclusion regarding Thomas's priority as these earlier studies: Craig L. Bartholomew, "Anthracite Iron," in *Proceedings of the Canal History and Technology Symposium*, vol. 3. Ed. Lance E. Metz (Easton, Penna.: Center for Canal History and Technology, 1984), 13–52.

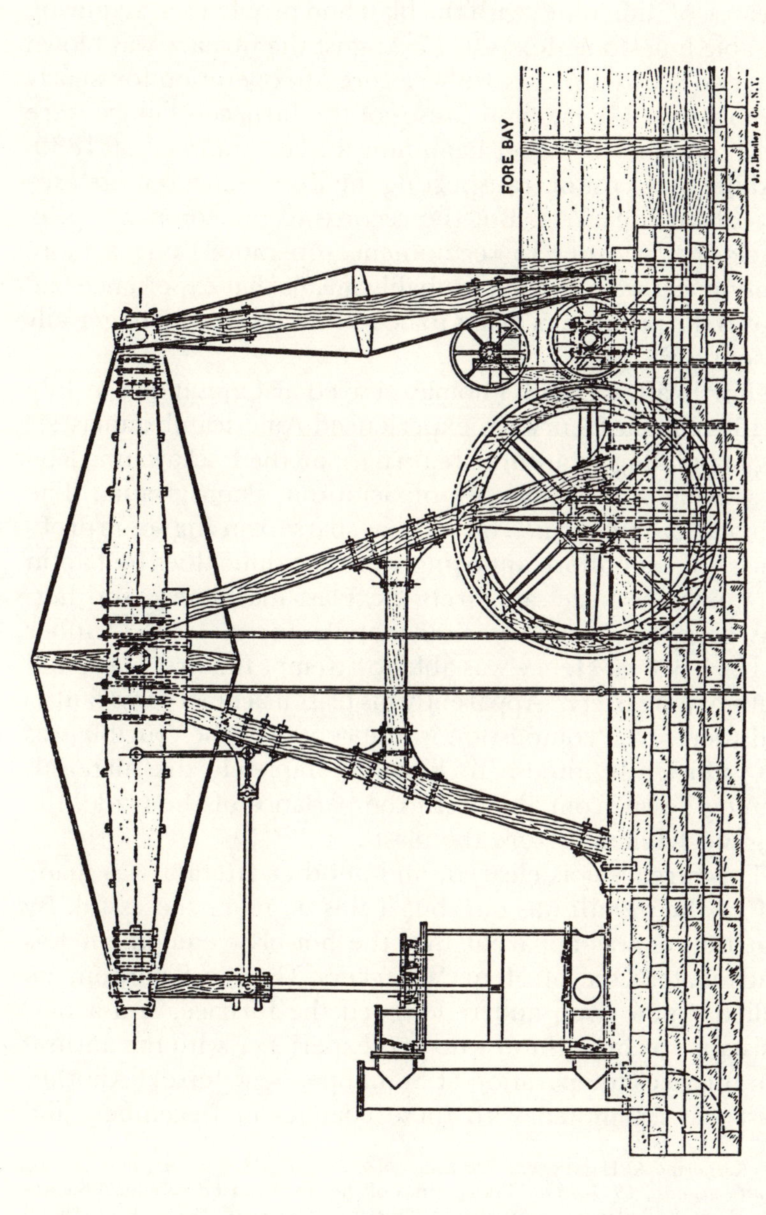

Fig. 15. Water-power blowing engine, Crane Iron Works, Catasauqua, Pennsylvania. Erected 1839–40. Vertical longitudinal section.

coke was used). The furnace produced modest quantities of adequate-quality pig iron. The furnace's performance varied because of difficulties with the blast and problems in acquiring suitable limestone flux. On 17 August the furnace was blown out, and was not successfully restored to operation for nearly seven years. The general causes of the furnace's failure were the economic recession beginning in late summer of 1839, and the high cost of transporting the iron from a remote area to seaboard markets. But the records of the company speak of closing in order to recommence operations on "a more economical scale" which probably means that experience had shown the promoters' plans to be technically and managerially unsound.[75]

Shortly after David Thomas arrived at Catasauqua in July 1840, William Henry, an experienced American ironmaster, began building an anthracite furnace on the Lackawanna River at what is now the city of Scranton, Pennsylvania. The financial backing which he had was sparse, causing some problems, but even more annoying were the difficulties he had in obtaining the necessary firebrick, blast machinery, and hot-blast apparatus from American manufacturers. It was October of 1841 before Henry was able to attempt the first blast, and that was a disaster.[76] Apparently his blast machinery was faulty, and therefore "combustion in the center of the crucible was not sufficiently intense to liquefy completely the materials coming down from above. In the parlance of the trade, the slag was 'chilling before the blast'. . . ."[77]

The furnace was cleared, and another attempt was made before the month was out, but it was no more successful. By that time it was apparent that the hot-blast equipment was inadequate. One of Henry's backers, George Scranton, installed a new oven and redesigned the furnace, and a new assistant was brought in who had experience with the anthracite furnaces in operation at Stanhope, New Jersey. Another blast was attempted with these changes in December, and

[75] Katherine A. Harvey, ed., *The Lonaconing Journals: The Founding of a Coal and Iron Community, 1837–1840.* Transactions of the American Philosophical Society, vol. 67, pt. 2 (Philadelphia: American Philosophical Society, 1977), 9, 14–15, 20, 22, 32, 40–44, 51–57, 66.

[76] Lewis, "Lackawanna Iron and Coal Company," 427–43.

[77] Ibid., 443.

although there was success for about two weeks, the furnace clogged again.[78]

The Lackawanna works were finally put in order by a Welsh ironmaster, John F. Davis, who was employed at one of the anthracite furnaces in the Danville area. He was hired in January 1842, and directed the repair of the furnace so that on the eighteenth of that month he was able to put in successful operation. From that point the furnace worked properly, although Davis modified the blast apparatus further because he found it inadequate.[79]

These episodes illustrate the intricate and obscure nature of smelting iron with coal, and the difficulty of transferring the technique from Britain to America. It is not surprising, then, that the proven skills of David Thomas were famous throughout the American iron industry for many years, and that the Lehigh Valley area was for thirty years the center of the coal-fueled iron industry.

But Thomas's fame came from more than the transfer of the anthracite iron process. He was also an innovator in furnace technique. His leadership in America was apparent immediately since the first furnace he built at Catasauqua was one of the largest and probably the best equipped in the United States. At the time it was built it had a higher blast pressure and a higher horsepower waterwheel, and blew more cubic feet of air per minute than any of the previous anthracite furnaces,[80] and probably more than any charcoal furnace.[81] A contemporary of Thomas stated that "with the erection of this furnace commenced the era of higher and larger furnaces and better blast machinery, with consequent improvements in yield and quality of iron produced."[82]

[78] Ibid., 443–46.

[79] Ibid., 446. Biographical data on Davis have not been found.

[80] Johnson, *Notes on the Use of Anthracite,* 28–29, 66.

[81] There is no definite authority for this, but it seems clear that anthracite furnaces were on the leading edge of innovation at this time.

[82] Firmstone, "Sketch," 155. Firmstone's statement, and the following discussion of innovations at the LCIC and Thomas Iron Company, indicate that a recent publication's assertion that the technology of the Pennsylvania iron industry in 1860 was much as it "had been more than a century earlier" is seriously flawed. Paul F. Paskoff, *Industrial Evolution: Organization, Structure and Growth of the Pennsylvania Iron Industry, 1750–1860* (Baltimore and London: Johns Hopkins University Press, 1983), 132.

Thomas remained an innovator in blast furnace size. For the Lehigh Crane Iron Company he built four more furnaces by 1850, all with a larger productive capacity, mostly due to larger size, than the first one.[83] In 1850 they were the most impressive group of furnaces in the United States.[84] These were exceeded only by those built in 1855 for his next employer, the Thomas Iron Company,[85] and although they were about the same size as the two previous ones, because of their innovative blast machinery and charging equipment they were regarded as "the model furnaces of America" in the early 1860s.[86]

As Thomas's furnaces grew larger, he made their blast machinery more powerful and their blast pressures greater. Although the pressures which were originally used at the LCIC works were only slightly above average, "about 1852 he introduced engines at his furnaces . . . which increased the pressure to double that which was customary in England."[87] His Thomas Iron Company works were reported in 1859 to be blown "at the extraordinary pressure of 8½ lbs. to the square inch [above atmospheric pressure]."[88] Thomas's example was not generally followed at first, but higher blast pressures became typical of advanced American blast furnaces in later years.[89]

Thomas was able to derive these powerful blasts not only from larger machinery—the blast cylinders of the first Thomas Iron Company furnaces were eighty-four inches in diameter—but also from different types of power.[90] Late in 1843 the directors of the LCIC began investigating the possibility of purchasing a water turbine for the works, and the following January made arrangements to have two made by Merrick

[83] Thomas, "Reminiscences," 916–25.

[84] See the list of anthracite furnaces in Pennsylvania, which had the vast majority of them: *American Railroad Journal* 23 (1850): 467.

[85] J. P. Lesley, *The Iron Manufacturer's Guide* (New York: John Wiley, 1859), 8–9; Bishop, *A History of American Manufactures,* 2: 576.

[86] Bishop, *A History of American Manufactures,* 2: 576.

[87] Swank, *History,* 328. Also see Thomas, "Reminiscences," 925, and Temin, *Iron and Steel,* 161.

[88] Lesley, *Iron Manufacturer's Guide,* 8.

[89] Temin, *Iron and Steel,* 161. See also John Fritz, *The Autobiography of John Fritz* (New York: John Wiley and Sons, 1912), 142–43.

[90] B. F. Fackenthal, Jr., *The Thomas Iron Company, 1854–1904* (Easton, Penna.: n.p., 1904), 11.

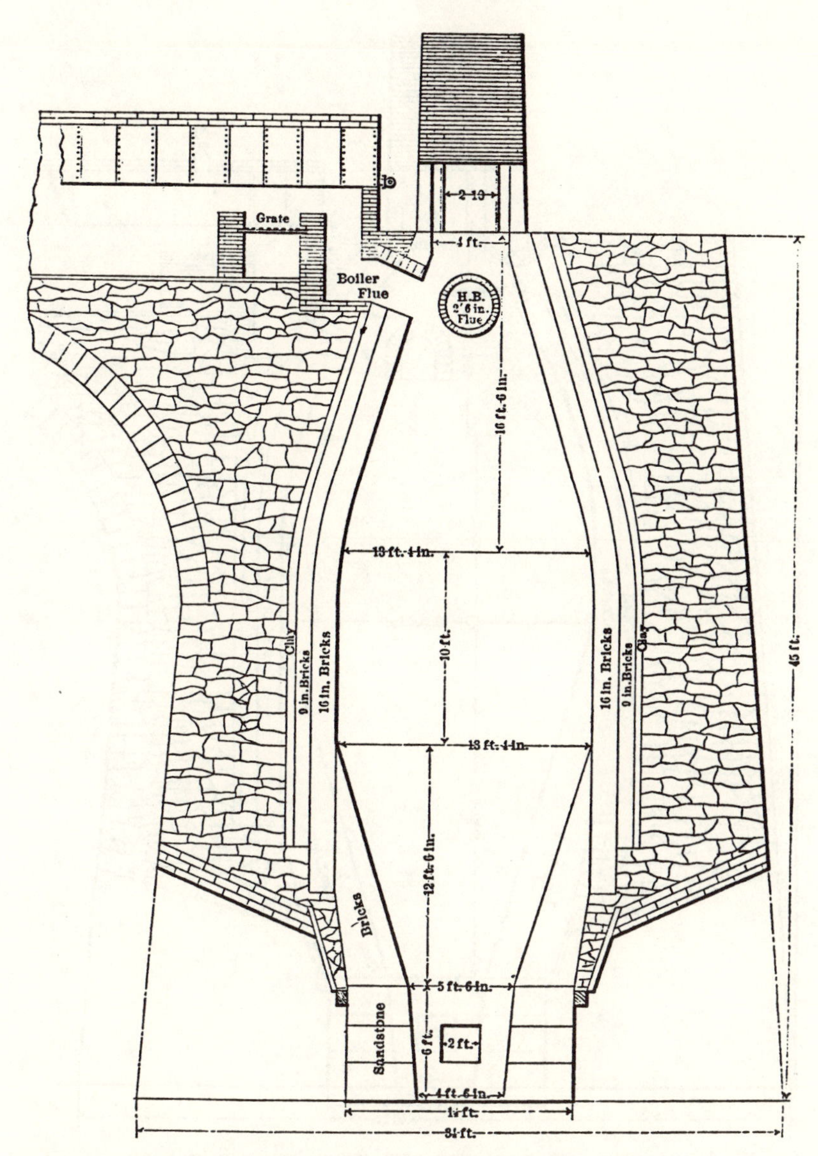

Fig. 16. Furnace No. 2, built at Catasauqua, Pennsylvania, in 1842.

and Towne.[91] They were installed sometime that year, although a company report stated that they used more water than the breast wheels (a second breast wheel had been built

[91] 4 December 1843, 1 January 1844, DM. For the history of the adoption of the turbine in the Philadelphia area, ca. 1843, see Louis C. Hunter, "Les Origines des turbines," *Revue d'histoire des sciences et de leur applications* 17 (1964): 216–18.

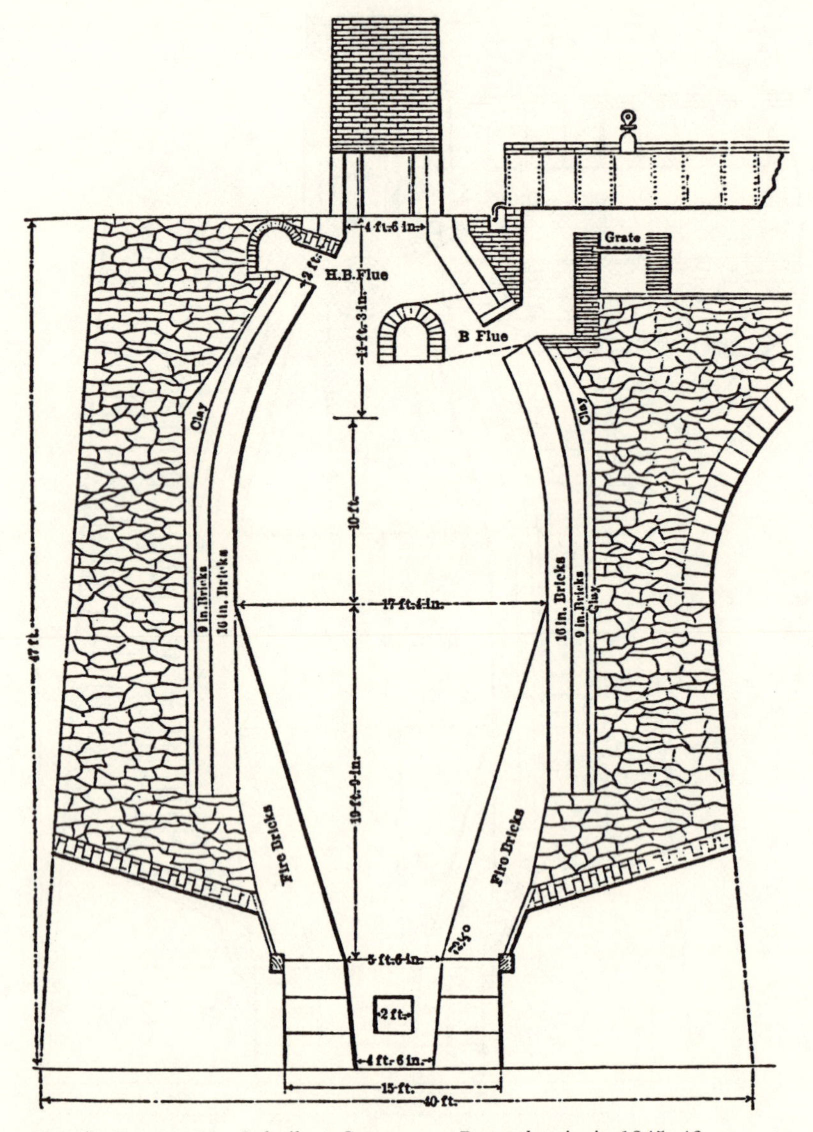

Fig. 17. Furnace No. 3, built at Catasauqua, Pennsylvania, in 1845–46.

to power the second furnace)[92] to do the same amount of work.[93] Apparently the company was satisfied with them, how-

[92] 14 November 1841, SM; cf. Thomas, "Reminiscences," 916–17.

[93] 11 November 1844, SM; cf. Thomas, "Reminiscences," 918, who incorrectly remembered the turbines being installed in 1842.

ever, because two years later all the water power was derived from them, and the waterwheels were used as auxiliaries.[94]

The third furnace, built in 1845–46, was blown with steam power.[95] In laying plans for the furnace David Thomas and a majority of the directors of the LCIC decided to use a steam engine with steam raised by the heat of the furnace, but at least one of the directors, Josiah White, had been adamantly opposed to it.[96] His two objections were:

> *First*—It is yet entirely a matter of *experiment* to obtain the power to drive the Engine from the Tunnel head of the Furnace, in *an anthracite* furnace, as no furnace using Anthracite Coal has been so driven. *Second*— . . . It would be manifestly imprudent to drive the furnace *air* by steam while we have abundance of water power on the canal free of cost. . . .[97]

White's first point suggests that David Thomas was an innovator in the operation of steam engines. Although they had been used for several years to blow the blast in some American furnaces,[98] the apparatus built for furnace number three was apparently the first at an anthracite furnace which heated the boilers with the furnace's escaping heat. Within a few years Frederick Overman, in his manual of ironmaking practice, *The Manufacture of Iron,* indicated that this technique was widely used.

> At most anthracite furnaces [steam is generated] . . . by putting the boilers on top of the furnace . . . . There is no reason whatever for employing water power in propelling blast machines at blast furnaces. There is abundance of waste heat for the generation of steam. The expense of erecting a steam engine will be found less, in most cases, than that incurred in the erection of a water wheel.[99]

While the engine built for the third LCIC furnace had a steam cylinder twenty-six inches in diameter with a six-foot

[94] 8 February 1847, SM.

[95] 27 October 1845, DM; 1 February 1846, SM.

[96] Thomas, "Reminiscences," 919, 921.

[97] Josiah White, 6 November 1845, to President and Board of the Lehigh Crane Iron Co., in 7 November 1845, DM.

[98] *American Railroad Journal* 23 (1850): 467; see n. 8 above.

[99] Frederick Overman, *The Manufacture of Iron in All Its Various Branches* (Philadelphia: Henry C. Baird, 1850), 180, 400.

stroke,[100] furnaces four and five, built in 1850, were blown by two engines working together which had steam cylinders thirty-four inches in diameter with nine-foot strokes.[101] These were the engines which in the early 1850s allowed Thomas to raise his blast pressure to the "extraordinary" level of eight and a half pounds per square inch. When the Thomas Iron Company furnaces were constructed they were powered by even larger engines.[102]

David Thomas was also associated with an early trial of a new method of refining iron by utilizing the gases escaping from the furnace stack. The LCIC was approached in September 1842 by C. E. Detmold, a German engineer living in America, with the proposition that the company utilize the gas process which had been developed and patented at Wasseralfingen in Germany by Wilhelm von Faber du Faur.[103] After some negotiation the company made an agreement with Detmold concerning the royalties to be paid to him as von Faber du Faur's agent, and then constructed the necessary apparatus.[104]

It was later described as "very similar to . . . a puddling-furnace," and was fueled by gases taken from the stack of the second furnace.[105] Puddling furnaces, or reverberatory furnaces, were commonly heated by coal, and used to convert pig iron into wrought or "refined" iron. There would have been a great saving of the expense of manufacturing wrought iron if the waste gases of the furnace could have been used for refining. After a trial in the latter half of 1843, however, the experiment was abandoned.[106] One difficulty was that the working of the chamber which removed the gases seriously damaged the furnace lining, which subsequently had to be replaced.[107] Another problem was that "after a rain . . . the

[100] Thomas, "Reminiscences," 924.

[101] Ibid., 925.

[102] Bishop, *A History of American Manufactures,* 2: 576; Fackenthal, *The Thomas Iron Company,* 11.

[103] 13 September 1842, DM; Swank, *History,* 328.

[104] 20 September 1842, 14 November 1842, 12 January 1843, DM; 22 November 1842, SM.

[105] Thomas, "Reminiscences," 918.

[106] Ibid., 919. Apparently there was an earlier trial which resulted in "the destruction of the gas chamber": 3 January 1843, DM.

[107] 11 November 1844, SM.

wet material going into the furnace so reduced the temperature of the gases . . . [that they were] insufficient to melt the iron."[108]

Although the Lehigh Crane Iron Company was not successful with using waste gas for refining iron, the idea continued to be widely discussed.[109] Detmold's device was used more successfully over the next twenty years to supply heat for hot-blast ovens and steam boilers.[110]

The LCIC's attempt to produce wrought iron should be looked at as one of the early attempts of the American iron industry at forward integration: that is, the completion at one site of two or more stages in the manufacture of a product. The LCIC for more than a year also seriously considered the construction of a rolling mill, probably to make rails, but the idea never had material results.[111] In contemplating the combination of making pig iron, refining it, and making it into rails, the LCIC was foreshadowing a development which even a decade later was barely underway in the United States.[112]

There was another field, however, in which the LCIC was technically far in advance of American practice. Walter Johnson described in 1841 the equipment which David Thomas had installed for handling the materials with which the furnace was charged.

> The *stock* at this furnace is very expeditiously elevated from the level of the base of the stack, by means of water pumped up by the blast wheel, into a cistern near the trunnel [sic] head, and which is thence allowed to flow alternately into two boxes of suitable dimensions, suspended by a chain passing over a pulley in such a manner, that the descent of one box filled with water, and bearing on its cover the empty barrows for stock, elevates

[108] Thomas, "Reminiscences," 919.

[109] J. Lawrence Smith, "Observations on the more recent researches in the Manufacture of Iron," *American Journal of Science*, 2nd ser., 2 (1846): 97–98; *American Railroad Journal* 22 (1849): 704.

[110] Swank, *History*, 328; Firmstone "Sketch," 156; Gale, *British Iron and Steel Industry*, 68. The LCIC probably used it for those purposes in later years, because in 1851 they paid royalties to Detmold: 9 February 1852, SM.

[111] 11 November 1844, SM; 7 April 1845, 4 May 1846, DM.

[112] A report in 1855 listed only six firms which had combined "the whole process of smelting and puddling": *American Railroad Journal* 28 (1855): 197. All these firms also had rolling mills: *American Railroad Journal* 27 (1854): 281. On the general subject of integration during this period see Temin, *Iron and Steel*, 90–94, 109–14, and Paskoff, *Industrial Evolution*, 117–18.

> the other box now emptied of water, but carrying up the barrows, loaded with ore, coal and limestone.[113]

What subsequent changes in the furnace-charging equipment were made is not known, but in 1847 the company built a special system of railroad tracks and trestles for unloading coal from canal boats. With that addition, the directors of the company stated that, "we have now nearly every desired convenience at our works, and unless it shall hereafter be thought expedient to lay down rail road tracks to our prominent mines we know not what else can be necessary."[114] David Thomas kept up his leadership in the area of raw materials handling at the Thomas Iron Company's works where, it was reported in 1864, "the materials [were] conveyed to the top of the stacks by atmospheric pressure."[115] The use of mechanical power was unusual, since at that time men or animals hauled the charge to the top of almost all American furnaces.[116]

In these innovations at the Lehigh Crane Iron Company David Thomas was the man who contributed the bulk of the technological "know-how," even though the directors of the company were men with substantial knowledge of the iron industry. Josiah White and Erskine Hazard had been, in the second decade of the nineteenth century, partners in a wire works outside of Philadelphia.[117] Nathan Trotter was a metal merchant who took a keen interest in developments in the American iron industry.[118] Other directors included the three Earp brothers, Thomas, George and Robert, who were also merchants in the metal trade, and one of whom, Robert, had some connection with a rolling mill.[119] These men sometimes gave instructions to David Thomas which expressed their technical judgment.[120]

[113] Johnson, *Notes on the Use of Anthracite,* 51–52. Cf. Thomas, "Reminiscences," 910.

[114] 14 February 1848, SM. A committee of the managers had consulted with Thomas concerning plans for a coal unloading apparatus early in 1846: 9 February 1846, DM. Cf. Thomas, "Reminiscences," 926.

[115] Bishop, *A History of American Manufactures,* 2: 576.

[116] Temin, *Iron and Steel,* 96n.

[117] Swank, *History,* 275.

[118] Elva Tooker, *Nathan Trotter, Philadelphia Merchant: 1787–1853* (Cambridge, Mass.: Harvard University Press, 1955), especially early chapters.

[119] 7 July 1845, DM; *Memoirs and Auto-Biography of Some of the Wealthy Citizens of Philadelphia* (Philadelphia: published by the booksellers, 1846), 20.

[120] E.g., 11 October 1842, 6 January 1845, DM.

Yet it is clear from the directors' minutes that David Thomas's opinion was highly valued and that he frequently had much to say about the company's course on technical matters. Invariably he was consulted by the directors on the construction of new equipment. For example, in planning for the second furnace they:

> Resolved That the President [Robert Earp] in conjunction with Erskine Hazard be requested to proceed forthwith in arranging and contracting for the construction of a Furnace of such dimensions as they and David Thomas may deem most advantageous to the interests of the company.[121]

Several years later when furnaces four and five were about to be built the directors requested that Thomas "attend a meeting of the Board at as early a day as convenient to submit to the Board his ideas and calculations as to the cost of said Furnaces."[122] Sometimes the board gave him complete responsibility for smaller items, as when they decided that "the replacing of the Water Wheel No. 1 be left to the discretion of David Thomas our Supt."[123] Thomas occasionally had the controlling voice in the company's decisions. The most striking example of this was recorded in the directors' minutes in the spring of 1844, "The Board met today to take into consideration the expediency of continuing the present pipe Contract but adjourned without coming to any definite conclusion awaiting further instruction from David Thomas."[124]

David Thomas was well compensated by the Lehigh Crane Iron Company for his work and skills. In the original agreement, signed with Erskine Hazard in Wales, Thomas was given an annual salary in the neighborhood of $800.[125] It was occasionally increased over the following years so that in 1847 he was paid $2,500, a substantial sum for the time.[126] The size of his salary allowed him in 1854 to become an investor in the Thomas Iron Company, which was formed by local capitalists and named after him. Its works were at Hokendauqua, about a mile up the Lehigh from Catasauqua. His

[121] 1 January 1841, DM.
[122] 27 November 1848, DM.
[123] 6 October 1845, DM.
[124] 27 May 1844, DM.
[125] Thomas, "Reminiscences," 905–906.
[126] 3 June 1844, 6 October 1845, 5 July 1847, DM.

Fig. 18. The Lehigh Crane Iron Company at Catasauqua, Pennsylvania, in 1860. Courtesy of Hagley Museum and Library.

Fig. 19. The Thomas Iron Works, Hokendauqua, Pennsylvania, in 1860. Courtesy of Hagley Museum and Library.

son Samuel was the superintendent, but Thomas exercised considerable influence in the management of the works.[127] He was also a large stockholder in and for a time president of the Catasauqua Manufacturing Company, an enterprise which had rolling mills at Catasauqua and Ferndale, Pennsylvania.[128] He was active as an investor or stockholder of the Carbon Iron Company, the Catasauqua and Fogelsville Railroad, the Lehigh Valley Railroad, and the National Bank of Catasauqua.[129] David Thomas was energetically managing his investments until near his death on 20 June 1882.

Perhaps the activity which Thomas most enjoyed in his last years was helping to found the American Institute of Mining Engineers (which included many in its membership from metallurgical trades), and his term as its first president in 1871. He was chosen because he was "the man whose name would do more than any other name to unite in support of our new enterprise the enthusiasm of science with the experience of practice."[130]

This honor demonstrated that David Thomas was recognized as a leader of his industry. To place his leadership in historical perspective, one must ask whether he was revered only as a symbol of the rapid rise of the iron industry or whether his practices and innovations were sources of major changes in American blast furnace technique? Although the evidence is indirect, it tends to suggest that Thomas's practices and innovations diffused to much of the American iron industry and affected it significantly.

The major method of diffusion must have been through other ironmasters visiting and inspecting the LCIC works at Catasauqua. At least in the first year, the company's managers actively promoted visits, probably because of their simultaneous interest in promoting consumption of the Lehigh Coal and Navigation Company's anthracite.[131]

[127] Roberts, "David Thomas," 9; John W. Jordan et al., *The Lehigh Valley* (New York: Lewis Publishing Company, 1905), 21; Lesley, *Iron Manufacturer's Guide*, 8–9; *An Act to Incorporate the Thomas Iron Company* (New York: W. E. & J. Sibell, 1855), 1; Fackenthal, *The Thomas Iron Company*, 7–8, 29.

[128] Roberts, "David Thomas," 9; Letters, Catasauqua Manufacturing Company, ca. 1868–73, Nathan Trotter Collection.

[129] Roberts, "David Thomas," 9; Jordan, *The Lehigh Valley*, 21.

[130] *Transactions of the American Institute of Mining Engineers* 11 (1883): 15.

[131] *Report of the Lehigh Coal and Navigation Company* (1840), 28; *The Miners' Journal*,

Thomas himself may have played an active role in the diffusion. Although his story is not fully corroborated by other sources, David Thomas claimed in later years that he was responsible for much of the success of the Pioneer furnace in Pottsville. That furnace had a widely publicized ninety-days' blast with anthracite late in 1839 before its machinery proved inadequate.[132] Thomas stated that the man who built Pioneer furnace had urged him to come to Pottsville in the summer of 1839 and that he "visited him in August, 1839, and furnished him with plans of in-wall, bosh, hearth, etc., and continued to visit him about once a month until the furnace was completed. . . . Then I was so engaged that I could not remain with him long enough to put it in blast."[133] Thereafter David Thomas's services may have been more closely guarded by his company, although in at least one instance a company was given permission "to make enquiries of David Thomas respecting the construction of Anthracite furnaces."[134]

In some instances men infused with David Thomas's ideas and skills must have left the LCIC to go to other anthracite ironworks,[135] although there is firm evidence for only one of his sons doing so. Samuel Thomas was trained under his father and then in 1848 built a furnace for the New Jersey Iron Company at Boonton, New Jersey.[136] He became superintendent of the Thomas Iron Company in 1854 and president in 1864, and three years later organized the Lock Ridge Iron Company and built its first furnace.[137] Another of David Thomas's sons, John, became superintendent of the Lehigh Crane Iron Company after his father's retirement.[138]

The ultimate impact of David Thomas's skill may lie in the connection of the anthracite and bituminous iron industries, the latter of which became the dominant branch of American

1 August 1840. Yates, "Discovery," 221, states that "the operation at Catasauqua became the principal model which other ironmasters sought to copy and improve on."

[132] See ns. 14 and 15 above.

[133] Thomas to Lynn, *Guide-Book.*

[134] 20 August 1847, DM.

[135] Darwin H. Stapleton, "The Diffusion of Anthracite Iron Technology: The Case of Lancaster County," *Pennsylvania History* 45 (April 1978): 147–57.

[136] Jordan et al., *The Lehigh Valley,* 22–23. This was the same company that asked for his father's advice in the previous year: 20 August 1847, DM.

[137] Jordan et al., *The Lehigh Valley,* 24.

[138] Bishop, *A History of American Manufactures,* 2: 576.

ironmaking after the Civil War.[139] A number of British ironworkers were important in the founding of the American bituminous industry, and direct transfer from Britain may have been the major source of technical skill in that branch.[140] The anthracite iron industry was, however, the first viable modern iron technology in the United States, and it pioneered in the use of many devices which were later adopted by the bituminous branch. The hot blast, more powerful blasts, steam engines, steam raised from the waste heat of the blast furnace, and waste gases used for steam boilers and hot-blast ovens were widespread in the anthracite iron businesses before the upsurge of the bituminous iron industry in the 1850s and 1860s.

Significant figures such as John Fritz moved from the anthracite iron industry to the bituminous iron industry, and several bituminous ironmasters trained under David Thomas.[141] An ironmaster trained at the Lehigh Crane Iron Company blew in the first coke furnace at Chattanooga, Tennessee, and a former employee of the Thomas Iron Company designed two of the early furnaces at Birmingham, Alabama.[142] William R. Jones (Captain Bill Jones), who had his apprenticeship at the Lehigh Crane Iron Company, became famous as the innovative superintendent of the J. Edgar Thomson Works of the Carnegie Steel Company. It should be noted that the art of "hand-driving" a furnace to obtain maximum production was probably begun by David Thomas, but did not become standard until Jones promoted it at the Thomson Works in the 1870s and 1880s.[143]

Even though the full extent of David Thomas's influence is difficult to assess, the story of his transfer of ironmaking technology from Wales to the United States is one of the clearest examples of the transfer of technology in the nine-

[139] Temin, *Iron and Steel*, 200, 266.

[140] Swank, *History*, 278–79, 283.

[141] Fritz, *Autobiography*.

[142] Ethel Armes, *The Story of Coal and Iron in Alabama* (Birmingham, Ala.: Birmingham Chamber of Commerce, 1910), 175–77, 354.

[143] S.v., "Jones, William Richard," in Allen Johnson and Dumas Malone, eds., *Dictionary of American Biography*, 20 vols. (New York: Charles Scribner's Sons, 1927–36); Temin, *Iron and Steel*, 157; Robert C. Allen, "The Peculiar Productivity History of American Blast Furnaces, 1840–1913," *Journal of Economic History* 37 (September 1977): 615.

teenth century. His skills, rooted in the British heritage of a coal-using technology, allowed him to play a major role in solving the anthracite coal problem at Yniscedwyn. He was brought to the United States by men who were concerned with the same problem in America, and he had an active part in the direction of the Lehigh Crane Iron Company, the company that they formed. Because of his innovative genius that company and the Thomas Iron Company which he helped to found remained leaders in the American iron industry long after the original transfer of Thomas's process proved to be a success. Thomas's influence was subsequently felt in the bituminous iron industry, and in that form a part of his legacy remains with us today.

Selected Bibliography

Adelman, Ira. *Theories of Economic Growth and Development.* Stanford: Stanford University Press, 1961.

Arbuckle, Robert D. *Pennsylvania Speculator and Patriot: The Entrepreneurial John Nicholson, 1757–1800.* University Park and London: The Pennsylvania State University Press, 1975.

Artz, Frederick B. *The Development of Technical Education in France, 1500–1850.* Cambridge, Mass.: SHOT and M.I.T. Press, 1966.

Basalla, George. "The Spread of Western Science." *Science* 156 (1967): 611–22.

Berthoff, Rowland Tappan. *British Immigrants in Industrial America, 1790–1950.* Cambridge, Mass.: Harvard University Press, 1953.

Bining, Arthur Cecil. *Pennsylvania Iron Manufacture in the Eighteenth Century.* Publications of the Pennsylvania Historical Commission, vol. 4. Harrisburg: Pennsylvania Historical Commission, 1938.

Bishop, J. Leander. *A History of American Manufactures.* 2 vols. Philadelphia: Edward Young and Co., 1861.

Bowen, Ele, ed. *The Coal Regions of Pennsylvania.* Pottsville, Penna.: E. N. Carvalho and Co., 1848.

Boyer, Nathalie Robinson. *A Virginia Gentleman and His Family.* Philadelphia: printed for the author, 1939.

Bridenbaugh, Carl. *The Colonial Craftsman.* Chicago and London: University of Chicago Press, 1961.

Brittain, James E. "The International Diffusion of Electrical Power Technology." *Journal of Economic History* 34 (1974): 108–21.

Brown, Robert R. "Pioneer Locomotives of North America." *Railway and Locomotive Historical Society, Bulletin* 101 (1959): 7–76. (Note the corrections to this article in *Bulletin* 103.)

Brzyski, Anthony Joseph. "The Lehigh Canal and Its Effect on the Economic Development of the Region Through Which it Passed." Ph.D. dissertation, New York University, 1957.

Calhoun, David Hovey. *The American Civil Engineer: Origins and Conflict.* Cambridge, Mass.: The Technology Press of M.I.T. Press, 1960.

Cameron, Rondo. "The Diffusion of Technology as a Problem in Economic History." *Economic Geography* 51 (1975): 217–30.

Carlson, Robert E. "British Railroads and Engineers and the Beginnings of American Railroad Development." *Business History Review* 34 (1960): 137–49.

Carroll, Charles F. *The Timber Economy of New England.* Providence: Brown University Press, 1973.

Centre National de la Recherche Scientifique. *L'Aquisition des techniques par les pays non-initiateurs.* Paris: Editions du Centre National de la Recherche Scientifique, 1973.

Chandler, Alfred D., Jr. "Anthracite Coal and the Beginnings of the

Industrial Revolution in the United States." *Business History Review* 46 (1972): 141–81.

Cippola, Carlo M. "The Diffusion of Innovations in Early Modern Europe." *Comparative Studies in Society and History* 14 (1974): 46–52.

Condit, Carl W. *American Building: Materials and Techniques From the First Colonial Settlements to the Present.* The Chicago History of Civilization. Chicago: University of Chicago Press, 1968.

De la Rivière, R. Dujarric. *E.-I. du Pont de Nemours: Elève de Lavoisier.* Paris: Librairie des Champs-Elysées, 1954.

Du Pont, B. G., ed. *Life of Eleuthère Irénée du Pont.* Newark: University of Delaware Press, 1923.

Dunaway, Wayland Fuller. *History of the James River and Kanawha Company.* New York: Columbia University, 1922.

Durfee, William F. "The Development of American Industries Since Columbus: III. Iron-Smelting by Modern Methods." *Popular Science Monthly* 38 (1890–91): 449–66.

Duveen, Denis I. and Herbert S. Klickstein. "Benjamin Franklin (1706–1790) and Antoine Laurent Lavoisier (1743–1794): Part II. Joint Investigations." *Annals of Science* 11 (1955): 271–308.

Fackenthal, B. F., Jr. *The Thomas Iron Company, 1854–1904.* [Easton, Penna.]: n.p., [1904].

Ferguson, Eugene S., ed. *Early Engineering Reminiscences (1815–1840) of George Escol Sellers.* Washington, D.C.: Smithsonian Institution, 1965.

———. "The American-ness of American Technology." *Technology and Culture* 20 (1979): 3–24.

Firmstone, William. "Sketch of Early Anthracite Furnaces." *Transactions of the American Institute of Mining Engineers* 3 (1874–75): 152–56.

Fitzpatrick, John C., ed. *The Writings of George Washington.* 39 vols. Washington: Government Printing Office, 1931–44.

Geise, John. "What is a Railway?" *Technology and Culture* 1 (1959): 68–76.

Gibson, George H. "Labor Piracy on the Brandywine." *Labor History* 8 (1967): 175–82.

Gray, Ralph D. "Philadelphia and the Chesapeake and Delaware Canal, 1769–1823." *Pennsylvania Magazine of History and Biography* 84 (1960): 401–23.

Grimaux, Edouard, ed., *Oeuvres de Lavoisier.* 6 vols. Paris: Imprimerie Nationale, 1862–93.

Glen, Robert. "Industrial Wayfarers: Benjamin Franklin and a Case of Machine Smuggling in the 1780s." *Business History* 23 (1981): 308–26.

Goulet, Denis. *The Uncertain Promise: Value Conflicts in Technology Transfer.* IDOC/International Documentation, no. 74, September/October 1976. New York: IDOC/North America, 1977.

Gruber, William H. and Donald G. Marquis, eds. *Factors in the Transfer of Technology.* Cambridge, Mass.: M.I.T. Press, 1969.

Hancock, Harold B. and Norman B. Wilkinson. "Joshua Gilpin: An American Manufacturer in England and Wales, 1795–1801." *Transactions of the Newcomen Society* 32 (1959–60): 15–28; 33 (1960–61): 57–66.

Hare, Jay V. *History of the Reading.* Philadelphia: John Henry Strock, 1966.

Harris, John R. "Saint Gobain and Ravenhead." In *Britain and Her World, 1750–1914: Essays in Honour of W. O. Henderson,* Barrie M. Ratcliffe, ed., pp. 27–70. Manchester: Manchester University Press, 1975.

Hart, George. "History of the Locomotives of the Reading Company." *Railway and Locomotive Historical Society Bulletin* 67 (1946): 1–119.

Heaton, Herbert. "The Industrial Immigrant in the United States, 1783–1812." *Proceedings of the American Philosophical Society* 95 (1951): 519–27.

Herskovits, Melville J. "The Processes of Cultural Change." In *The Science of Man in the World Crisis,* Ralph Linton, ed., pp. 143–70. New York: Columbia University Press, 1945.

Hindle, Brooke. *David Rittenhouse.* Princeton: Princeton University Press, 1964.

———. *Technology in Early America.* Chapel Hill: University of North Carolina Press, 1966.

A History of the Lehigh Coal and Navigation Company. Philadelphia: William S. Young, 1840.

Hunter, Louis C. *Waterpower in the Century of the Steam Engine. A History of Industrial Power in the United States, 1780–1930.* Vol. 1. Charlottesville: University of Virginia Press, 1979.

Hyde, Charles K. *Technological Change in the British Iron Industry, 1700–1870.* Princeton: Princeton University Press, 1977.

———. "The Adoption of the Hot Blast by the British Iron Industry: A Reinterpretation." *Explorations in Economic History* 10 (1973): 281–93.

Jennings, Francis. *The Invasion of America: Indians, Colonialism, and the Cant of Conquest.* New York: Norton, 1976.

Jeremy, David J. "British Textile Technology Transmission to the United States: The Philadelphia Region Experience, 1770–1820." *Business History Review* 47 (1973): 24–52.

———. "Damming the Flood: British Government Efforts to Check the Outflow of Technicians and Machinery, 1780–1843." *Business History Review* 51 (1977): 1–34.

———. "Immigrant Textile Machine Makers Along the Brandywine, 1810–1820." *Textile History* 13 (1982): 225–48.

———. *Transatlantic Industrial Revolution: The Diffusion of Textile Technologies Between Britain and America, 1790–1830s.* Cambridge, Mass.: M.I.T. Press, 1981.

Jeremy, David and Anthony F. C. Wallace. "William Pollard and the Arkwright Patents." *William and Mary Quarterly,* 3rd ser., 34 (1977): 404–25.

Johnson, Keach. "The Genesis of the Baltimore Ironworks." *Journal of Southern History* 19 (1953): 157–79.

Kirby, Richard Shelton. "William Weston and His Contribution to Early American Engineering." *Transactions of the Newcomen Society* 16 (1935–36): 111–27.

Labaree, Leonard W. and William B. Willcox et al., eds. *The Papers of*

Benjamin Franklin. 25 vols. in progress. New Haven and London: Yale University Press, 1959–.

Leavitt, Thomas W., ed. *The Hollingsworth Letters: Technical Change in the Textile Industry, 1826–1837.* Cambridge, Mass.: SHOT and M.I.T. Press, 1969.

"Letters of Moncure Robinson to his Father, John Robinson, of Richmond, Va., Clerk of Henrico Court." *William and Mary Quarterly,* 2nd ser., 8 (1928): 71–95, 143–56; 9 (1929): 13–33.

Lewis, W. David. "The Early History of the Lackawanna Iron and Coal Company: A Study in Technological Adaption." *Pennsylvania Magazine of History and Biography* 96 (1972): 424–68.

Livingwood, James Weston. *The Philadelphia–Baltimore Trade Rivalry, 1780–1860.* Harrisburg: Pennsylvania Historical and Museum Commission, 1947.

McKie, Douglas. *Antoine Lavoisier: The Father of Modern Chemistry.* Philadelphia: J. B. Lippincott Company, 1936.

Mayr, Otto and Robert C. Post, eds. *Yankee Enterprise: The Rise of the American System of Manufactures.* Washington: Smithsonian Institution Press, 1981.

Minchinton, W. E. "The Place of Brecknock in the Industrialisation of South Wales." *Brycheiniog* 7 (1961): 1–70.

Mordecai, John B. *A Brief History of the Richmond, Fredericksburg and Potomac Railroad.* Richmond: Old Dominion Press, 1941.

Morison, Elting E. *From Know-How to Nowhere: The Development of American Technology.* New York and Scarborough: New American Library, 1977.

Mulholland, James A. *A History of Metals in Colonial America.* University: University of Alabama Press, 1981.

Multhauf, Robert P. "The French Crash Program for Saltpeter Production." *Technology and Culture* 12 (1971): 163–81.

North, Douglass C. *The Economic Growth of the United States, 1790–1860.* New York: W. W. Norton & Co., 1966.

[Osborne, Richard Boyse]. *Sketch of the Professional Biography of Moncure Robinson, Civil Engineer.* Philadelphia: J. B. Lippincott Co., 1889.

Palmer, Arlene M. "A Philadelphia Glasshouse, 1794–1797." *Journal of Glass Studies* 21 (1979): 102–14.

———. "Glass Production in Eighteenth-Century America: The Wistarburgh Enterprise." *Winterthur Portfolio* 11 (1976): 75–101.

Parker, William N. "Economic Development in Historical Perspective." *Economic Development and Cultural Change* 10 (1961): 1–7.

Passer, Harold. *The Electrical Manufacturers, 1875–1900.* Cambridge, Mass.: Harvard University Press, 1953.

Penn, Theodore Z. "The Introduction of Calico Cylinder Printing in America: A Case Study in the Transmission of Technology." In *Technological Innovation and the Decorative Arts,* Ian M. G. Quimby and Polly Anne Earl, eds., Winterthur Conference Report 1973, pp. 235–51. Charlottesville: University of Virginia Press, 1974.

Percival, Arthur. "The Faversham Gunpowder Industry." *Industrial Archeology* 5 (1968): 1–42.

Pursell, Carroll W., Jr., *Early Stationary Steam Engines in America: A Study in the Migration of a Technology.* Washington: Smithsonian Institution Press, 1969.

———, ed. *Technology in America: A History of Individuals and Ideas.* Cambridge, Mass.: M.I.T. Press, 1981.

———. "Thomas Digges and William Pearce: An Example of the Transit of Technology." *William and Mary Quarterly,* 3rd ser., 21 (1964): 551–60.

Roberts, Christopher. *The Middlesex Canal, 1793–1860.* Cambridge, Mass.: Harvard University Press, 1938.

Roberts, Ed. *David Thomas: The Father of the Anthracite Iron Trade.* Westfield, Wales: The Red Dragon, [1883].

Roberts, Solomon W. *Memoir of Josiah White: Written as a Chapter in the History of the Lehigh Valley.* Easton, Penna.: Bixler & Corwin, 1860.

———. "The Early History of the Lehigh Coal and Navigation Company." *Railway World* 1 (1875): 297–99.

Robinson, Eric H. "The Early Diffusion of Steam Power." *Journal of Economic History* 34 (1974): 91–107.

Rogers, Everett M., with F. Floyd Shoemaker. *Communication of Innovations: A Cross-Cultural Approach.* New York: The Free Press, 1971.

Roland, Alex. "Bushnell's Submarine: American Original or European Import?" *Technology and Culture* 18 (1977): 157–74.

Rolt, L. T. C. *George and Robert Stephenson: The Railway Revolution.* London: Longmans, Green and Co., 1960.

Rosenberg, Nathan. *Perspectives on Technology.* Cambridge: Cambridge University Press, 1976.

———. *Technology and American Economic Growth.* New York: Harper and Row, 1972.

Rubin, Julius. *Canal or Railroad? Imitation and Innovation in Response to the Erie Canal in Philadelphia, Baltimore, and Boston.* Transactions of the American Philosophical Society, new ser., vol. 51, pt. 7. Philadelphia: American Philosophical Society, 1961.

Scoville, Warren C. "Minority Migrations and the Diffusion of Technology." *Journal of Economic History* 11 (1951): 347–60.

Scranton, Philip. "An Immigrant Family and Industrial Enterprises: Sevill Schofield and the Philadelphia Textile Manufacture, 1845–1900." *Pennsylvania Magazine of History and Biography* 106 (1982): 365–92.

Shaw, Ronald E. *Erie Water West: A History of the Erie Canal, 1792–1854.* Lexington: University of Kentucky Press, 1966.

Shelton, Cynthia. "Labor and Capitol in the Early Period of Manufacturing: The Failure of John Nicholson's Manufacturing Complex, 1793–1797." *Pennsylvania Magazine of History and Biography* 106 (1982): 341–64.

Shyrock, Richard H. "British Versus German Traditions in Colonial Agriculture." *Mississippi Valley Historical Review* 26 (1939): 39–54.

Stapleton, Darwin H. "The Diffusion of Anthracite Iron Technology: The Case of Lancaster County." *Pennsylvania History* 45 (1978): 147–57.

———, ed. *The Engineering Drawings of Benjamin Henry Latrobe.* The Papers of Benjamin Henry Latrobe, ser. 2, vol. 1. New Haven and London: Yale University Press, 1980.

———. "The Origin of American Railroad Technology, 1825–1840." *Railroad History* 139 (1978): 65–77.

Stapleton, Darwin H. and Thomas C. Guider. "The Transfer and Diffusion of British Technology: Benjamin Henry Latrobe and the Chesapeake and Delaware Canal." *Delaware History* 16 (1976): 127–38.

Strassman, W. Paul. *Technological Change and Economic Development.* Ithaca: Cornell University Press, 1968.

Swank, James M. *History of the Manufacture of Iron in All Ages.* Philadelphia: published by the author, 1884.

Temin, Peter. *Iron and Steel in Nineteenth-Century America: An Economic Inquiry.* Cambridge, Mass.: M.I.T. Press, 1964.

Thomas, Samuel. "Reminiscences of the Early Anthracite-Iron Industry." *Transactions of the American Institute of Mining Engineers* 29 (1899): 901–28.

Thompson, Mack. "Causes and Circumstances of the Du Pont Family's Emigration." *French Historical Studies* 6 (1969): 58–77.

Turner, Charles W. "The Early Railroad Movement in Virginia." *Virginia Magazine of History and Biography* 55 (1947): 350–71.

Upton, Dell. "Traditional Timber Framing." In *Material Culture of the Wooden Age,* Brooke Hindle, ed., pp. 35–93. Tarrytown, N.Y.: Sleepy Hollow Press, 1981.

Uselding, Paul J. "Henry Burden and the Question of Anglo-American Technological Transfer in the Nineteenth Century." *Journal of Economic History* 30 (1970): 312–37.

———. "Studies of Technology in Economic History." *Research in Economic History,* supplement 1 (1977): 159–219.

Van Gelder, Arthur Pine and Hugo Schlatter. *History of the Explosives Industry in America.* New York: Columbia University Press, 1927.

Von Oeynhausen, C. and H. von Dechen. *Railways in England: 1826 and 1827.* Charles E. Lee and K. R. Gilbert, eds., E. A. Forward, trans. Cambridge: The Newcomen Society, 1971.

Wallace, Anthony F. C. *Rockdale: The Growth of an American Village in the Early Industrial Revolution.* New York: Alfred A. Knopf, 1978.

White, John H., Jr. *American Locomotives.* Baltimore: The Johns Hopkins Press, 1968.

Wilkins, Mira. "The Role of Private Business in the International Diffusion of Technology." *Journal of Economic History* 34 (1974): 166–88.

Wilkinson, Norman B. "Brandywine Borrowings from European Technology." *Technology and Culture* 4 (1963): 1–13.

Wilson, W. Hasell. *Reminiscences of a Railroad Engineer.* Philadelphia: Railway World Publishing Company, 1896.

INDEX